Volume 3 · 1983 · Number 3

Acta Biotechnologica

Journal of microbial, biochemical and bioanalogous technology

Akademie-Verlag
Berlin

THEORIA CUM PRAXI

ISSN 0138-4988
Acta Biotechnol., Berlin 3 (1983) 3, 193–296
EVP 30,– M

Instructions to Authors

1. Only original papers that have not previously been published will be accepted.
   Manuscripts may be submitted in English, German or Russian but always in duplicate. The name of the institute (inclusive full address) from which the manuscript originates should be stated below the author's name(s). The authors are responsible for the contents of their contributions.

2. Original papers should not consist of more than 20 typewritten pages (2 lined-spacing), including references, tables and figures; short original communications may have a maximum of 6 typewritten pages.

3. Each paper should be preceded by two summaries, in each case one in english.

4. Latin names of species as well as passages to be printed in italics for greater emphasis should be signed with weved line. Please use only units and symbols of the SI-system.

5. Tables may be used to effect a reduction of the text or to make assimilation easier. They should be numbered consecutively throughout and supplied with a brief heading. They shouldn't be written among the text but on separate sheets.

6. Number and size of illustrations should be limited to the essential minimum, they should be numbered consecutively and brought on separate sheets.
   Line drawings including graphs and diagrams should be drawn in black ink. Half-tone illustrations should be supplied as white glossy prints. Figure legends are to be typed in sequence on a separate sheet.
   Authors name is to be written on to the rear of each sheet.

7. References, listed at the end of the contribution, should contain only works cited in the text. They should be numbered in the order in which they are at first mentioned in the text. Please give surnames and initials of all authors, name of journal abbreviated according to "Chemical Abstracts — List of Periodicals", volume number, year of publication, issue number or month, first page number; books are to be cited with full title, edition, volume number, page number, place of publication, publisher and year of publication.

8. Notes on the text may be given as footnotes on the relevant page.

9. 50 reprints are free of charge.
   Additional reprints may be ordered.

10. The author receives 2 galley proofs for correction.
    They are sent back to the redaction
    (Dr. Dimter, DDR - 7050 Leipzig, Permoserstr. 15)
    as soon as possible.

Acta Biotechnologica

Journal of microbial, biochemical and bioanalogous technology

Edited at
Institute of Technical Chemistry of the Academy of Sciences of the G.D.R.; Leipzig
and
Institute of Technical Microbiology; Berlin
by
M. Ringpfeil, Leipzig and G. Vetterlein, Berlin

Editorial Board:

M. E. Beker, Riga
H. W. Blanch, Berkeley
S. Fukui, Kyoto
H. G. Gyllenberg, Helsinki
G. Hamer, Zurich
J. Holló, Budapest
M. V. Iwanow, Pushchino
F. Jung, Berlin
H. W. D. Katinger, Vienna
K. A. Kalunjanz, Moscow
J. M. Lebeault, Compiegne
D. Meyer, Leipzig
P. Mohr, Berlin
P. Moschinski, Lodz
A. Moser, Graz
M. D. Nicu, Bukarest
Chr. Panajotov, Sofia
L. D. Phai, Hanoi
H. Sahm, Jülich
W. Scheler, Berlin
R. Schulze, Köthen
B. Sikyta, Prague
G. K. Skrjabin, Moscow
M. A. Urrutia, Habana
J. E. Zajic, El Paso

Redaction: L. Dimter, Leipzig

Volume 3

1983

Number 3

AKADEMIE-VERLAG · BERLIN

"Acta Biotechnologica" publishes reviews, original papers, short communications and reports out of the whole area of biotechnology.
The journal shall promote the foundation of biotechnology as a new, homogeneous scientific field.
According to biotechnology are microbial technology, biochemical technology and technology of synthesyzing and applying of bioanalogous reaction systems.
The technological character of the journal is guaranteed thereby that microbial, biochemical, chemical and physical contributions must show definitely the technological relation.

Terms of subscription for the journal "Acta Biotechnologica"
Orders can be sent

- **in the GDR:** to Postzeitungsvertrieb or to Akademie-Verlag, DDR - 1086 Berlin, Leipziger Str. 3—4;
- **in the other socialist countries:** to a book-shop for foreign languages literature or to the competent news-distributing agency;
- **in the FRG and Berlin (West):** to a book-shop or to the wholesale distributing agency Kunst und Wissen, Erich Bieber OHG, Wilhelmstr. 4—6, D - 7000 Stuttgart 1;
- **in the other Western European countries:** to Kunst und Wissen, Erich Bieber GmbH Dufourstr. 51, CH - 8008 Zürich;
- **in other countries:** to the international book- and journal-selling trade, to Buchexport, Volkseigener Außenhandelsbetrieb der DDR, DDR - 7010 Leipzig, Postfach 160, or to Akademie-Verlag, DDR - 1086 Berlin, Leipziger Str. 3—4.

Acta Biotechnologica

Herausgeber: Institut für technische Chemie der AdW,
DDR - 7050 Leipzig, Permoserstr. 15
(Direktor: Prof. Dr. Manfred Ringpfeil) und
Institut für Technische Mikrobiologie
DDR - 1017 Berlin; Alt-Stralau 62
(Direktor: Dipl. Ing. G. Vetterlein)

Verlag: Akademie-Verlag, DDR - 1086 Berlin, Leipziger Straße 3—4;
Fernruf: 2236221 und 2236229; Telex-Nr.: 114420; Bank: Staatsbank der DDR, Berlin,
Konto-Nr.: 6836-26-20712.

Redaktion: Dr. Lothar Dimter (Chefredakteur),
Martina Bechstedt (Redakteur),
DDR - 7050 Leipzig, Permoserstr. 15; Tel.: 2392255.

Veröffentlicht unter der Lizenznummer 1671 des Presseamtes beim Vorsitzenden des Ministerrates der Deutschen Demokratischen Republik.

Gesamtherstellung: VEB Druckhaus „Maxim Gorki", DDR - 7400 Altenburg.

Erscheinungsweise: Die Zeitschrift „Acta Biotechnologica" erscheint jährlich in einem Band mit 4 Heften. Bezugspreis eines Bandes 120,— M zuzüglich Versandspesen; Preis je Heft 30,— M.

Bestellnummer dieses Heftes: 1094/3/3.

Urheberrecht: Alle Rechte vorbehalten, insbesondere der Übersetzung. Kein Teil dieser Zeitschrift darf in irgendeiner Form — durch Photokopie, Mikrofilm oder irgendein anderes Verfahren — ohne schriftliche Genehmigung des Verlages reproduziert werden. — All rights reserved (including those of translations into foreign languages). No part of this issue may be reproduced in any form, by photoprint, microfilm or any other means, without written permission from the publishers.

© 1983 by Akademie-Verlag Berlin. Printed in the German Democratic Republic.

AN (EDV) 42133

Acta Biotechnol. 3 (1983) 3, 195–216

Formal Macroapproach to Bioprocessing-Modeling with Analogies

A. Moser

Institute for Biotechnology, Microbiology and Waste Technology, Technical University Graz, Austria, A-8010, Schlögelgasse 9

Summary

A new approach for the investigations of biotechnical processes is presented in this article on the basis of an integrating strategy, incorporating a joint analysis of biological and engineering aspects. Thereby, the experimental work can be carried out on the macroscopic level, taking into account the behaviour of bioprocesses from measurements in the liquid phase. Mathematical modeling uses the concept of analogies instead of speculative mechanisms. A predictive character of these "formalkinetics" supplementary to the descriptive nature is achieved with the aid of deductive research methodology. This general theoretical concept is illustrated in the special case of modeling the diauxic growth of yeast. Formalkinetics based on analogies are shown to be not only able to be successfully applied to elucidate bioprocessing but also to form a more realistic and simple approach than pure mechanistic investigations needed for the realization of bioprocessing in technical scale.

Zusammenfassung

Der Artikel stellt eine neue Bearbeitungsweise für biotechnische Prozesse vor, die sich auf Basis einer gesamtheitlichen Betrachtungsweise zwischen biologischen und verfahrenstechnischen Aspekten ergibt. Demnach sind experimentelle Untersuchungen auf dem makroskopisch in der Flüssigphase pseudohomogen beobachtbaren Niveau durchzuführen. Mathematische Modellierungen erfolgen dann mit Hilfe von Analogien anstelle rein spekulativ mechanischer Vorstellungen. Der beschreibende Charakter dieser „Formalkinetik" kann auf voraussagende Eigenschaften erweitert werden, indem die deduktive Methodologie in der wissenschaftlichen Bearbeitung befolgt wird. Dieses allgemein theoretische Konzept wird am Beispiel der Hefediauxie im Detail dargelegt. Analogien zeigen sich demnach nicht nur für die wissenschaftliche Durchdringung von Bioprozessen geeignet, sondern representieren eine im Vergleich zu rein mechanistischen Ansätzen realistischere und einfachere Bearbeitungsmethode, wie sie für die technische Realisierung von Bioprozessen auch notwendig ist.

Introduction

Bioprocessing as a "skilled craft" in the beginning of this new field of technology resulted mainly from trial and error methods. Later on a pragmatic approach was used in the past where dominantly engineering data with poor biological background formed the basis of industrial realization, which approach is still used in present activities.

Biotechnology is devided into two separated parts: the field of bio-sciences (biology and biochemistry) and technical aspects (process technology with all engineering sciences). These technical activities were mainly carried out by chemical engineers who were thinking and working in analogies to chemical processes. Biotechnology, however, is thought to be an interdisciplinary field, where microbiology, biochemistry and process technology contribute. So, biotechnology is defined to be only the integrated activity of the three constituent disciplines. Nevertheless the research situation in the past has characteristically been that each discipline applied own methods with a strong tendency to treat the microbiological and biochemical aspects of metabolism in isolation from the chemical engineering aspects of reactors with their physical transport phenomena. Furthermore, the internal metabolic fluxes are normally estimated indirectly from measurements of external metabolic exchange rates due to the fact that in vivo-measurements are very difficult to realize.

The in vitro-investigations of isolated parts of metabolism which are normally predominantly examined in biochemical and biological laboratories, however, are inadequate for the quantification of the highly coordinated and organized networks of intact systems of living cells. As a consequence, the application of engineering data to actual bioprocessing situations is a difficult task and often misleading as well as to apply biological data from isolated experiments to technical scale-bioreactors. The complexity of biotechnical problems forced scientists and engineers to elaborate biological and physical phenomena in isolation. This situation can be accepted as long as there is no need to synthesize both aspects which, however, is the case when bioprocesses must be carried out reproducibly in industrial scale. The development of bioprocesses for technical purpose can no longer be carried out following trial and error methods because the main part of new processes in industry must be economically realized in less than 10 years. A systematic approach to bioprocess design which must be able to fulfill this need, therefore, is to be formed on the basis of an integrating strategy [33, 34] where integration meens more than a summing up [22]. This article demonstrates the general procedure of this strategy, describing and illustrating with some case studies the basic working principles according to a macroscopic approach where balances of essential process variables are considered together with kinetic modeling using a formal kinetic level elaborated with the aid of analogies. The main aim is to show that a certain level of simplification in quantification and modeling is not only a necessity but a freely chosen approach without lost of significant information.

Basic working principles

Experimental verification nowadays is still restricted by the fact that only poor measuring methods are known for bioprocess control. Even if computers can be used for solving kinetic equations, the main problem still is to identify and estimate model parameters from experimental runs. Following the macro approach presented the level of research work in bioprocessing can be reduced in details and accuracy but not in the level of a 'joint' point of view between the three disciplines. The gap between fundamental and engineering sciences can only be bridged over by this integrating strategy where both phenomena of biology and physics are taken into account simultaneously but on a reduced number of details.

An engineering approach to bioprocessing follows four working principles in analysis, outlined in detail elsewhere [29, 33]:

1. to quantify, 2. to simplify, 3. to separate biological and physical phenomena, 4. to model both parts and recombine them in process design.

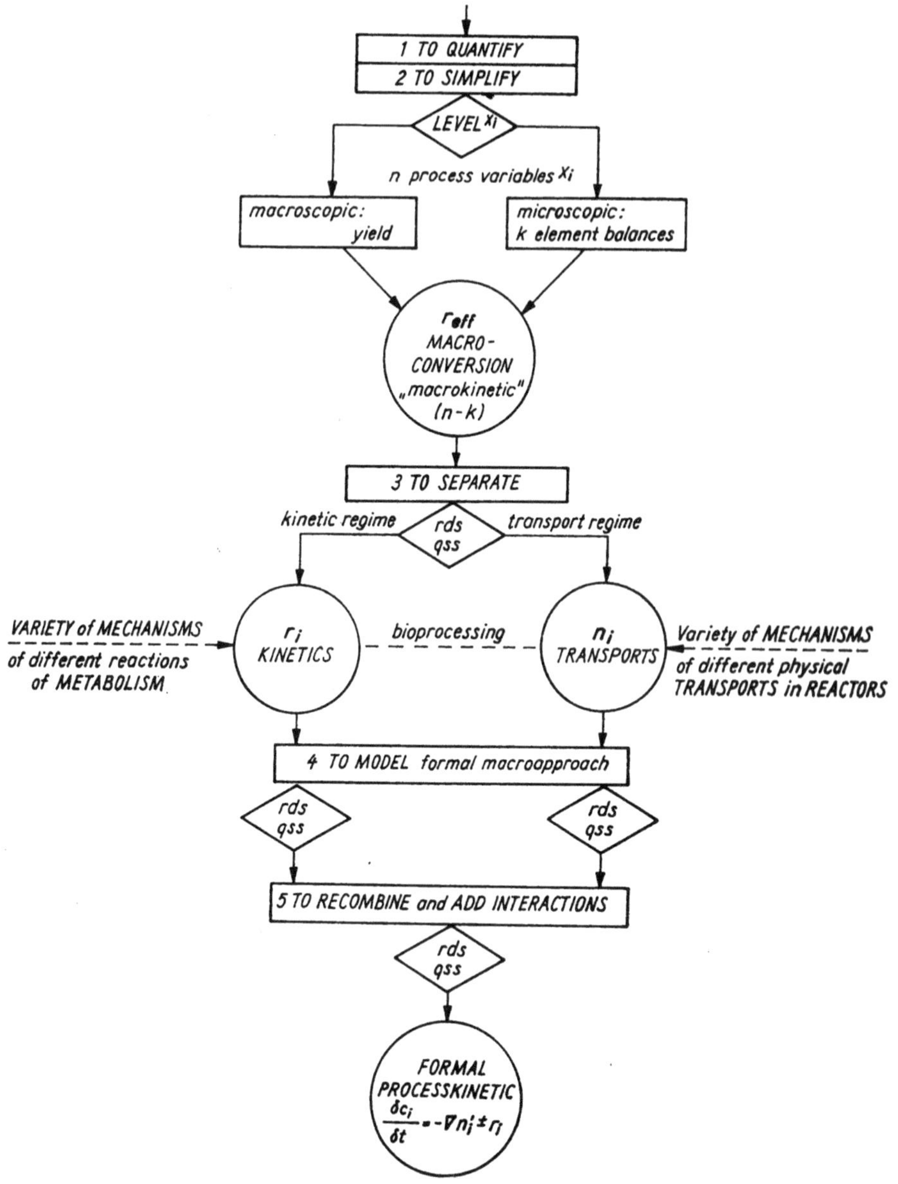

Fig. 1. Engineering approach to bioprocessing on macroscopic level according to the integrating strategy using five working principles and the basic concepts of rate-determining-step (rds) and quasi-steady-state (qss) to formulate process kinetics

This approach is illustrated with the scheme on Fig. 1, showing the systematic procedure of process analysis and design. The variety of mechanisms of either metabolic reactions and physical transport in reactors dominating the 'horizontal' level of the variety of different processes in biotechnology are shown to be densified and generalized on the 'vertical' level of basic principles.

The problem of quantifying and simplifying bioprocessing is manifested when the level of process variables is chosen. A macroscopic approach operates often with a number n

of macroscopic variables conventionally measured, e.g. concentration of biomass (x), substrates (s_i), oxygen (o), products (p), carbon dioxid (c) and heat of reaction (h_v) resp. temperature (T). Stoichiometry in this case is replaced by the use of the concept of yield coefficients $y_{i/j}$. A more microscopic approach uses balancing of the essential elements like carbon C, oxygen O, hydrogen H, nitrogen N and eventually sulfur S and phosphorus P. In this case of k elementary balances only a number of $(n-k)$ kinetic equations are necessary for a complete description of the process [17, 18, 49]. It becomes clear therefore that the number of independent kinetic equations to be postulated and verified by experiments cannot be chosen at will [15].
In any case of approach, measurements of concentration changes in bioreactors are to be considered as effective rates (r_{eff}) because these concentration changes are the result of reaction rates interwoven with transport phenomena (so called "macrokinetics"). This definition is taken in analogy to chemical engineering [59] and contains the interactions between reaction kinetics and external plus internal transport limitations. The influence of micromixing in the liquid-phase is excluded by definition, but is hard to be separated in experimental runs. Therefore, it seems to be advantageous for bioprocessing to include this influence in the macrokinetics [29, 34]. As a consequence of the kinetics interwoven with transports the main tool in process analysis is to separate both phenomena of biology and physics. Tests can be carried out in some analogy to chemical processing [18, 53], adapted to bioprocessing as a "test of pseudohomogeneity" by the author [30, 33]. With the aid of the concepts of rate-determining-step ("rds") and quasistationarity ("qss") a "kinetic regime" can be distinguished from a "transport regime", so that biokinetics can be investigated without limitation by physical transports and vice versa.
In the same manner the following two basic principles of modeling and recombining both phenomena can be applied using again the concepts of "rds" and "qss". This is discussed in more detail later. The result of this strategy is the formal description of process behaviour ("formal process kinetics") which contain both effects of metabolism and transports in combination and with all interactions. The level of these process kinetics is given by the chosen macroscopic methods and the property is characterized by the mathematically formulated structure of the model equations ("formal kinetics").

The macroscopic approach

It was shown that in agreement with the integrating strategy a joint analysis of bioprocesses can and must be reduced to a certain simplified level in both parts of biological and physical phenomena. This macroscopic description is based on the measurement of macroscopic variables as indicated before. This approach is used since long time in the analysis of physical systems and in reaction engineering [59, 11, 52] and was recently consequently applied to bioprocessing by Roels [49].
The formal macroapproach [27], reviewed in this paper, is strongly related to the macroscopic principle. The general background is the application of the conservation law of mass etc. [50] by formulating the balance equations of interest. In the case of bioprocessing, the basic situation is schematically shown on Fig. 2, where the process behaviour is quantified with the aid of the changes of concentrations of the components X, S_i, O, P, C, H_V.

The overall change

$$\frac{\partial c_i}{\partial t} = -\nabla n_i' \pm r_i \tag{1}$$

is given by interactions between true reaction kinetics (r_i, kmol/m^3 · h) and physical transports, where n_i' = flux of mass through area (kmol/m^2 · h) and ∇ = spatial variation (∂/∂_z etc.). Transport can be caused by conductive, convective and/or interfacial transport mechanisms. This basic eqn. 1 quantifies the macro-conversion and is defined as "macrokinetics" when the interactions are ignored and measurements are considered only as reaction rate, r_{eff}.

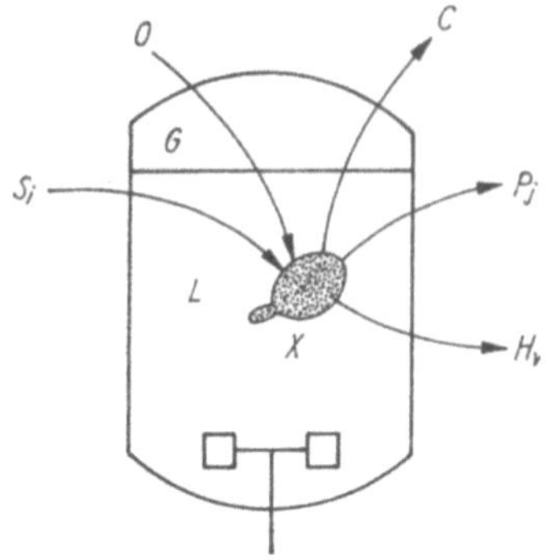

Fig. 2. Schematic representation of the macroscopic behaviour of bioprocesses on the basis of the pseudohomogeneous observable process variables in the liquid phase (L): biomass X, substrates S_i, oxygen O, products P, carbon dioxyd C and volumetric heat H_v

From eqn. 1 the true reaction rates in the case of different pseudohomogeneous reactors can be derived. These definitions are useful in process analysis and design [33]. Different modes of operation are to be distinguished:

1. stationary with $\partial c/\partial t = 0$
2. instationary with $\partial c/\partial t \neq 0$
3. ideal reactors with neglectable conduction so that only convection must be taken into account

$$\frac{\partial c_i}{\partial t} = \text{div}\,(v \cdot c_i) \pm r_i \tag{2}$$

with v = flow velocity

4. non-ideal reactors with convection and conduction

$$\frac{\partial c_i}{\partial t} = -\text{div}\,(\text{grad}\, c_i) + \text{div}\,(v \cdot c_i) \pm r_i. \tag{3}$$

In the case of ideal continuous stirred tanks (cSTR) eqn. 2 must be integrated over all 3 directions of space so that a term of flow rate F (m^3/h) appears. Varying the flow rate, these definitions can be generalized to variable-volume reactors e.g. fed batch cultures. These reactors have "lumped" parameters while discontinuous stirred tanks (dcSTR) and plug flow reactors (cPFR) have distributed parameters. Both terms of transports n_i and reactions r_i in eqn. 1 must be treated according to the macroscopic method. The variety and complexity of bioreactors and biokinetics must be reduced to simplified but adequate systems, i.e. models.

Types of bioreactor models are:

1. heterogeneous or homogeneous
2. stationary or instationary
3. distributed (integral or with gradients) or lumped (differential or gradientless) parameter reactor models.

The first consequence when working with the macroapproach, therefore, is to reduce any experimental situation of a bioreactor to one of these fundamental types ("bioreactor model").

Only significant transports are taken into consideration:

- macromixing e.g. residence time distribution (RTD) with limiting cases of cSTR and cPFR as representatives for lumped and distributed parameter situations
- micromixing e.g. mixing time (t_m). This measurement is useful for defining lumped-parameter-reactors (STR-types)
- O_2-transfer (OTR) through G/L- and L/S-phase and particle diameter of biocatalytic phase, flocs or films (d_P) in order to characterize the heterogeneous case.

The stationary or instationary situation is realized by a constant or varying flow rate. A more precise quantification of the influence of this mode of operation is achieved with the aid of characteristic time constants or rate constants of the environment (t_{env}) and the bioreaction (t_r). This will be discussed later.
Furthermore, the variety and complexity of biological reactions must be reduced to a simplified but adequate scheme ("biokinetic model"). Types of biokinetic models are, once again, described as

1. heterogeneous or homogeneous
2. stationary or instationary
3. distributed or lumped parameter models.

These cases are strongly related to the bioreactor model chosen.
The simpliest case of biokinetic models is well understood nowadays, represented by the homogeneous, stationary lumped parameter situation, normally elucidated in a cSTR-configuration. The basic behaviour of cell metabolisms, schematically shown in Fig. 2, can be quantified with the aid of the corresponding rates of growth (μ), substrat consumptions (σ_S, σ_O) and productions (π, π_C, π_{H_v}).
For all these cases exist a large number of models summarized in recently written papers and books [5, 43, 50, 33]. A systematic classification is possible distinguishing between

1. numerical curve fitting, using arbitrary mathematical functions with "anonymous" parameters, which do not contain any biological significance or
2. "formal kinetics" with fully empirical background or with "partially proved" mechanisms.

This will be explained in more detail in the next session. Both methods are used to describe the phenomena observed on the basis of μ, σ and π. Microbial growth and substrat consumption follow generally the laws of stimulation and saturation kinetics, where at higher values of the influencing variable (S, T, etc) inhibition and toxicity occur with the consequence that a "biological optimum" exists in a certain narrow range of conditions. A typical phenomenon for living systems is the endogeneous metabolism and "biological inertia", which must be considered in modeling. Microbial production are treated as growth associated or growth non-associated on phenomenological level of investigation. This empirical fact can be described with the aid of some mechanistic derivations, where μ and π are dependent on simple enzyme kinetics. In this way the appearance of the empirical association with growth can be explained with some deeper biological background [4].
The more complicated cases of biokinetic models (heterogeneous, instationary case etc.) are quite well established in the case of heterogeneous processing [1, 33] but still at the beginning of examination in the case of dynamic processing. Here, the basic working principles discussed should be very useful. All cases of interactions between transports in environment and reaction in the cells are thought to obey the same law. The concept of "rds" and "qss" can be used to simplify the system [35, 36].

The specific properties of the formal macroapproach, reviewed in this paper, is summarized in Tab. 1. The result of this type of convenient bioprocess analysis is balance equations of the significant process variables with kinetic terms for the special bioreactor situation.
The process behaviour is represented in diagrams, where the specific rates of growth, consumptions and productions are shown in dependence from time (dcSTR) or mean residence time resp. dilution rate (cSTR; cPFR). From these plots, the kinetic models must be identified and the model parameters estimated according to the formal kinetic approach.

Tab. 1. Specific properties of the formal macroapproaches using two different levels of process-variables

macroscopic observable concentrations: x, s_i, o, p_j, c, h_v	level of process-variables	microanalysis of significant elements: C, N, O, H; S, P
mainly irreversible	Thermodynamics	mainly irreversible
$Y_{i,j}$ (yield-concept)	Stoichiometry	$\nu_{i,j}$ (stoich. coefficients)
empirical phenomena	formal kinetics	mechanistic background
using ANALOGIES ("if-then") as first hypothesis, must be compared to experiments according to the deductive method, "adaption-parameters" "black box"	methodology	using speculative or partially proven mechanisms as basis for the derivation of kinetic equations containing parameters with more biological interpretation difficult to be verified experimentally "grey box"
mainly descriptive, predictive in the range of "systematic validity" e.g. "kinetic similarity"	models	descriptive and predictive

Formal kinetic modeling using analogies

The formal kinetic approach uses mathematically formulated functions in analogy to other similar situations. These functions are made adequate to singular situations by adapting their parameters to experiments. These formal kinetics can be constructed fully empirical taking known model functions as analogies or by using some know-how about the mechanism to sustain the choice of specific model function of reactions. It is, however, important to state that in both fields of kinetic modeling, the formal character of the approach still remains, also when some feeling exists that a partially proven mechanism is to be preferred. One should not forget, however, that the elaboration of "true kinetics" is impossible because of all general difficulties we have in bioprocessing (and also with chemical processes) which do not allow a "causal" description.
All we can do is to adapt the formal kinetics to reality of the process. This fundamental and essential procedure of constructing "simplified but adequate" models, is illustrated on Fig. 3 [27, 28, 29, 33]. The procedure of modeling according to the formal macroapproach must be carried out in both situations of modeling, i.e. the bioreactor and the biokinetics. An adapted procedure includes steps of induction ("to conclude from a singular situation to general trends", which is the case when starting with experiments)

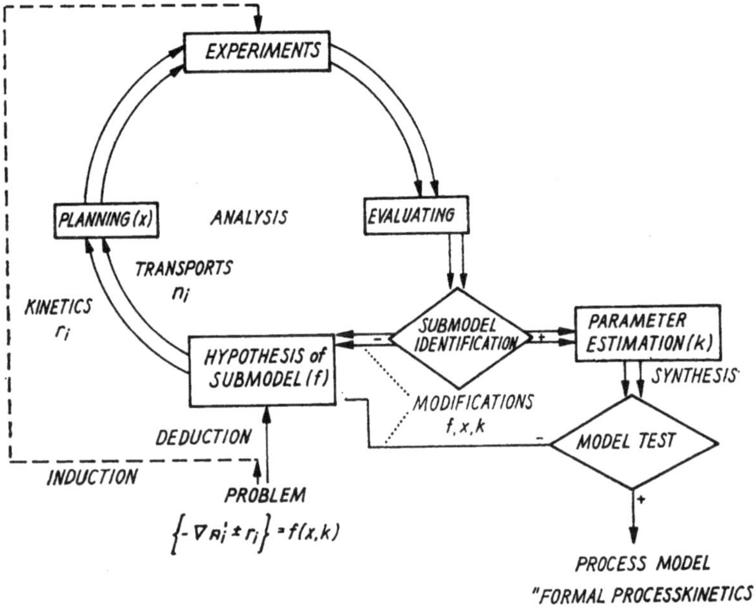

Fig. 3. Procedure of adaptive mathematical modeling of bioprocess-kinetics following a combination of deductive and inductive steps of research

and steps of deduction when starting with a formulated hypothesis according to general knowledge. Starting with an analogy as a first hypothesis the experiments must be designed, so that the experimental runs are able to give all data necessary for model identification.

Step by step, after model identification, the model parameters must be estimated according to one of the methods of differential or integral method of evaluation. Finally the separately determined models for transports and kinetics must be recombined and tested if there exist some interactions.

The most important fact, which simultaneously seems to be the main draw-back at first view, is the hypothesis taken at the beginning in "analogy to similar situations". Is this step still scientific work? Is it not so that by adapting parameters to experiments "true" relations between theory and practise are lost? This question can only be answered with the aid of the theories of scientific research. According to the modern theory of deduction [44, 45] causality is hard to achieve because we always observe singular facts. We are not justified from a logical point of view to infer from singular to general statements. Any conclusion drawn in this way may always turn out to be false once, no matter how often it was confirmed in the past. Theories are always "fallible" and the only way to circumvent this possibility is to compare hypothesis with experiments. In the case of coincidence in between normal deviations one has found a "workable solution". The erroneous belief that science eventually leads to the certainty of a definite explanation carries with it the implication that scientists think it is a grave scientific mistake to have operated with hypothesis or theories that eventually are falsified. Research effectiveness is thought to be strongly related to a proper and systematic research methodology [32]. Due to the fact that theories are fallible and a causalnexus not directly achievable the author concludes that the only logical way in making research is to operate with analogies. All we can do is to apply the principle of "if ..., then ...",

which implicitely stresses that we have to compare hypothesis and experiments (cf. Fig. 3). Using this research methodology the formal character of all approaches becomes evident. The advantage of formal approach is that we are able to transfer on this basis the mathematical function to any other arbitrary situation. The disadvantage of this concept with the potential property of generalizable functions is the fact that they cannot give any answer to the question "why".

Saturation-type-kinetics

One example is demonstrated directly here to clarify the meanings of formal kinetics. The equation 4

$$r = \frac{k_1 \cdot c}{1 + k_2 \cdot c} \tag{4}$$

describing saturation-type kinetics can be used for different types of reactions.

a) for homogeneous chemical reaction with varying order of reaction n e.g. from 0 to 1. In the case of $n = 0$ is

$$r = k_1/k_2 = r_{\max} \tag{5}$$

and with $n = 1$

$$r = k_1 \cdot c \tag{6}$$

b) for the pseudohomogeneous approach to heterogeneous reactions e.g. chemical catalysis. Here we have the well known LANGMUIR-equation

$$r = k \cdot \frac{K \cdot c}{1 + K \cdot c} \tag{7}$$

with

$$k_1 = k \cdot K \tag{8}$$

$$k_2 = K \tag{9}$$

K = equilibrium constant of sorption and

$$r = k \cdot \Theta \tag{10}$$

where Θ = fraction of covered surface

c) for a pseudohomogeneous description of enzymatic reactions arises the MICHAELIS-MENTEN equation:

$$v = v_{\max} \frac{s}{K_m + s} \tag{11}$$

which has the same form as eqn. 4 ($v = r$) with the identity-equation

$$K_m = \frac{1}{K}. \tag{12}$$

Furthermore $K_m = 1/k_2$ and $v_{\max} = k_1/k_2$

d) for pseudohomogeneous microbial growth and substrate consumption the MONOD-equation is valid:

$$\mu = \mu_{\max} \frac{s}{K_S + s} \tag{13a}$$

resp.

$$-\sigma = \sigma_{max} \frac{s}{K_S + s} \qquad (13b)$$

which are related to a yield concept (Y)

$$\sigma_{max} = \frac{1}{Y_{x/S}} \mu_{max} \qquad (14)$$

Comparison to eqn. 4 and 11 show that

$$K_S = \frac{1}{K} \qquad (15)$$

$$K_S \doteq K_m \qquad (16)$$

with

$$K_S = 1/k_2 \quad \text{and} \quad \sigma_{max} = k_1/k_2 \qquad (17a)$$

but

$$\sigma = v \cdot \frac{1}{x}. \qquad (17b)$$

Equation 17b becomes clear when we take into account that in the case of dcSTR:

$$\sigma = -\frac{ds}{dt} \cdot \frac{1}{x} \qquad (18)$$

but

$$v = -\frac{ds}{dt}. \qquad (19)$$

Enzyme kinetics as normal catalytic reactions occur with constant concentration of catalysts while microbial reactions are of autocatalytic nature so that the biomass concentration X as a measure for biocatalytic active mass is to be considered.
Eqn. 4, therefore, is the generalized form valid in all mentioned cases. This formal kinetic equation, however, cannot give any hint to a mechanism. On the other side Langmuir-kinetics (eqn. 8) and enzyme kinetics (eqn. 11) can be derived, postulating a mechanism, where molecules are adsorbed on the surface of the catalyst X before the reaction and desorbed afterwards:

$$S + X \rightleftharpoons \{SX\} \rightleftharpoons \{PX\} \rightleftharpoons P + X. \qquad (20)$$

It is interesting to note that in the case of enzyme kinetics a quite different approach according to Briggs and Haldane result in a similar analogous equation as according to the Michaelis-Menten-mechanism. This statement confirms the formal kinetic approach, operating with analogies.
In the same way the difference between enzyme kinetics and microbial Monod kinetics can be explained both resulting in an analogous form of equation (cf. eqn. 11 and 13). Even if the parameters K_m and K_S seem to be identical, one has to take care under which conditions the formal approach on the vertical level of principles can be transferred to the horizontal level of reality (cf. Fig. 1). The singular situation is quite different, because in the case of microbial cells, a multi-step-process must be included in a hypothetical mechanism (cf. Fig. 2). Monod kinetics and mainly the model parameter K_S contains not only the enzyme kinetic reaction step (hyperbolic function acc. eqn. 11) but also the adsorption step on the cell surface, which obeys also a hyperbolic function acc. eqn. 8

and also all transports through membranes on outer and inner surface, which again are enzyme catalized and obey eqn. 11.
The difficulty when applying the formal kinetic approach of MONOD equation is always the same, i.e. lack of a strong relation to any causality. Even when the experiments were made with e.g. obvious C-limitation it cannot be concluded that the causalnexus is C-limitation because at the same time transports or energy supply or any enzyme reaction can be limiting or even substrate and product can be inhibiting. Therefore we have to accept the formal character of eqn. 4 in the case of microbial kinetics by the knowledge that the magnitude of K_S can vary tremendously compared to the value of K_m [33] e.g. the value of K_S is directly a function of diameter of the cell due to cell-internal transports [10].
Finally, an important question concerns the "test of adequacy". Formal kinetics should be simple but adequate. Operating with iterative model building in Fig. 3, a hypothetical situation can arise, where an "unexpected" model function is also able to fit the data. The decision that a model is really adequate, therefore, can be easier made by "putting the model in jeopardy" i.e. to apply extreme concentrations. In this case the basic function often still can be used to represent the data, but only with varying model parameters. Therefore a test of adequacy is a test of parameter sensitivity, to show if values of the parameters are constant with varying conditions. This fact is demonstrated by discussing the appearance of a simple kinetic approach on the basis of arbitrary reaction order (cf. eqn, 5 and 6).
In literature some examples can be found where a good fit to experimental data in the case of substrate consumption can be achieved with the aid of reaction orders $n = 0, 1/2$ and 1.
Arbitrary reaction order concept: The case of $n = 0$ is described in activated sludge treatment, where the normal biokinetics are falsified by the phenomenon of biosorption by sludge [54, 33]. The reason for this effect is that values of COD represent the substrate concentration in the liquid phase ("elimination kinetics") while the BOD-value is related to the O_2-consumption which is caused by the substrate removal from the liquid phase due to substrate degradation in the solid phase. The substrate is firstly eliminated from the liquid phase and then it is degraded in the sludge. This know-how taken in analogy enables to elucidate the deviation from normal microbial kinetics.
Reaction order $n = 1$: This case is also found in waste water technology where simultaneously a variety of substrates is removed by the sludge but measurements are carried out by global methods like COD and BOD. It was shown [64, 65] that the reaction order $n = 1$ can be explained by this multi-substrate reaction.
Another case of $n = 1$ is described when operating with biofilm reactors [1, 7]. In the case of mass transfer limitation this empirical formal kinetics with $n = 1$ appear.
A similar case is analyzed in literature with a reaction order $n = 1/2$ [14], which type of kinetics is again the consequence of internal transport limitations.
An arbitrary reaction order is thought to appear in the case of multiple reactions where the single reactions obey expected kinetics (cf. eqn. 11). The overall reaction follows eqn. 21

$$r = k \cdot c^{1/n} \tag{21}$$

which is an analogon to the adsorption isotherm according to FREUNDLICH [12].
From these considerations it seems to be very advantageous to start model discrimination in bioprocessing with a differential analysis of eqn. 21 in order to estimate reaction order and to conclude from this order to some overlooked phenomena previously explained as r_{eff}.
Normally, operating with mathematical models, descriptive models are distinguished from predictive models. A descriptive model, where the microbial cells are considered

as a black box, is thought to give only information about the "how" of the system behaviour but not about the "why". This is only possible with predictive models, which are based on mechanisms [50]. Against this scientific statement, however, the following arguments can be found.
Predictive models, as previously defined, represent an explanatory structure and allow an extrapolation to other conditions due to the basic mechanism. The final subsystems, however, are always black boxes, so that this grey-box models are an optical illusion. They seem to be valid with more statistical determined probability. A closer analysis, however, shows that mechanism can never be proved to be certain. All we are able to do is, according to the deductive research methodology, to reject hypotheses or theories. The only difference, as already outlined, lies in the fact that formal kinetics from a purely empirical approach can be sustained by a mechanistic background (cf. tab. 1) which satisfies our feelings but it is never logically based. The relationship between reason and effect, i.e. a causalnexus, is never ascertained, so that we should be satisfied primarily to clarify the know-how i.e. using the principle of analogy ("if..., then...").
The application of this principle of analogy, however, is not a loss of information but a different approach. The microscopic structure of metabolism is, according to the macroscopic principle, always manifested in the macroscopic behaviour. This statement can be accepted more easily if we know that due to the "equifinality" of biological systems [60], as a consequence of the "flowing equilibrium" in open systems in living cells [38] the mathematical modeling becomes easier. Living systems are "equifinal" i.e. that they finally reach the same state at the end independent of initial conditions. Biological systems behave like standardized or unified systems where the process inside the "black box" is highly "organized". Chemical processes on the other hand are more chaotic and not so coordinated in their microscopic pathways of reaction network. This biophysical background forms the theoretical justification applying the formal macro-approach. It is therefore sufficient for the quantification of engineering purpose to "know, how" the system behaves, when conditions are changing. This is the principle of analogy ("if, then").
Analogy or homology means that different phenomena, even if they follow different mechanisms, their externally observed macroscopic behaviour exhibit similar properties with an identical mathematical type of function on the formal level of elaboration.
This world of working with analogies should be demonstrated in detail in the case of the diauxie of yeasts.

Case study: diauxic growth of yeast

The phenomenon of diauxic growth i. e. the sequential utilization of carbon sources by microbial cells is extensively examined in literature. The overall behaviour can be described with the aid of the following scheme

$$S_1 + O \xrightarrow{X} X + P + C + H_V \qquad (22a)$$

and (dashed arrow from P to S_2)

$$S_2 + O \xrightarrow{X} X + C + H_V \qquad (22b)$$

with $S_2 = P$ (ethanol)
The mathematical modeling of yeast-diauxie has been attempted from different levels of approach. 1) purely empirical models representing a phenomenological nonmechanistic attempt to quantify the effect of preferential utilization of a particular substrate [24, 6, 23, 27]. 2) empirical models on the basis of measurable physicochemical process variables of the culture like in the case 1 but with more "biological significance" by

incorporating some simplified mechanistic information of metabolism (e.g. ATP, inhibition of active transport systems) [40, 41, 42]. 3) more mechanistic descriptions of the control processes based on theoretical mechanisms of yeast metabolism, including terms to quantify catabolite repression, catabolite inhibition, induction of enzyme etc. on a genetic level [58, 25, 39, 3, 56]. Similar considerations of regulatory actions of glucose upon both utilization of ethanol and glucose catabolism with the known glucose- and O_2-effects (Pasteur-, Crabtree-effects) were used as basis for predicting optimal feed strategies in yeast technology [42, 65].
The choice of appropriate level of modeling according to the philosophy of modeling is dependent on the aim of investigations. The "know-how" is explored following the attempts with empirical models (case 1 or 2) while the "know-why" is elucidated following the concept of more mechanistic approaches. Considering the difficulties in experimental verifications, with poor analytical methods in situations where physical transports are always interacting with biokinetics must lead to accept some restrictions and

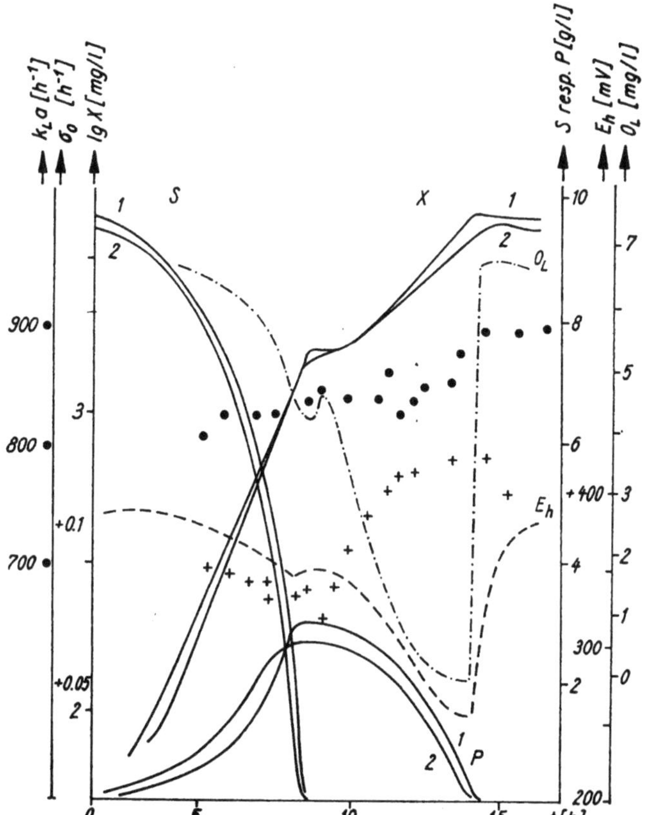

Fig. 4. Experimental data of the batch culture of *Saccharomyes cerevisiae* under aerobic conditions in a synthetic medium acc. Fiechter [8] with 1% glucose: Concentration/time plot of the macroscopic process variables X, S, P, O in the liquid phase together with measurements of the volumetric oxygen transferrate k_La, the redox potential E_h and the oxygen consumption rate σ_O.
The yeast cultivation is carried out in a conventional stirred tank bioreactor (curve 2) and in an agitated tubular reactor with liquid recycle ATF (curve 1) with a working volume of 25 l [26]

simplifications. From the engineering point of view, as it was already stated previously, we have to take into account the real situations in process technology e.g. in biological waste water treatment, where the biological oxygen demand (BOD) consists of complex mixtures of carbon sources, which can be utilized simultaneously and/or sequentially. Due to the variety of the behaviour of metabolism interacting with environment, it is emphasized by the author that nowadays the only attempt to be applied successfully is the formal macroapproach. With increasing knowledge of biochemical network of metabolic reactions, but a lack in quantifying its cell-internal fluxes, we cannot expect some significant contributions in the near future. Experimental runs of a batch culture of *Saccharomyces cerevisiae*, using a synthetic medium according to Fiechter [8] with glucose as limiting substrate, are shown on Fig. 4 and 5 with different initial concentra-

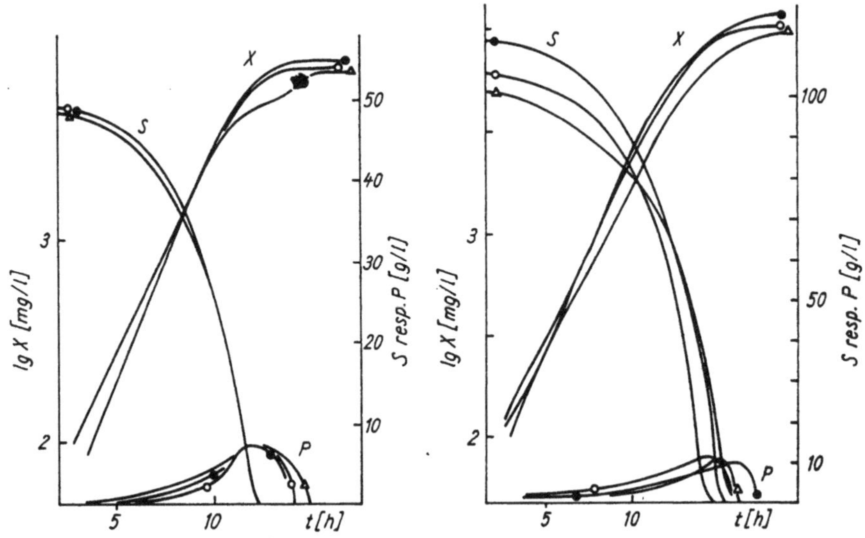

Fig. 5. Experimental data of the batch culture of *Saccharomyces cerevisiae* under aerobic conditions in a synthetic medium [8] with 5% (a) resp. 10% (b) glucose as limiting substrate. Compared are runs in the ATF with 240 rpm (•), 160 rpm (○) and in the stirred tank with 1200 rpm (△)

tions ($S_0 = 10$, 50, 100 g/l). The time dependence of the concentrations of biomass X, substrate S and ethanol P and oxygen O are compared, when operating with a conventional stirred tank (STR) and an agitated tubular reactor with recycle (ATF) at different rotational speeds [26]. Beyond the kinetic interpretations analyzed in detail elsewhere [28], it can be easily shown that the diauxic behaviour depends on S_0 and the mixing behaviour of the fluid phase in the bioreactors used.

The basis of a formal kinetic description of diauxic growth of yeast are the measureable process variables indicated on Fig. 2, which are quantified from Fig. 4 and 5 in the form of diagrams representing the specific rates of utilization (substrate σ, oxygen σ_0) and formation (growth μ, product π). This quantification plot (cf. Fig. 6) serves as background for the set up of mathematical models according to the adaptive procedure on Fig. 3. The diauxic growth is kinetically analyzed separately in phase I with glucose as substrate and in phase II with ethanol as substrate.

An Eadie-Hofstee-plot of the first growth phase is shown on Fig. 7. As a first working hypothesis the simple Monod equation seems to be adequate beyond the deviations in

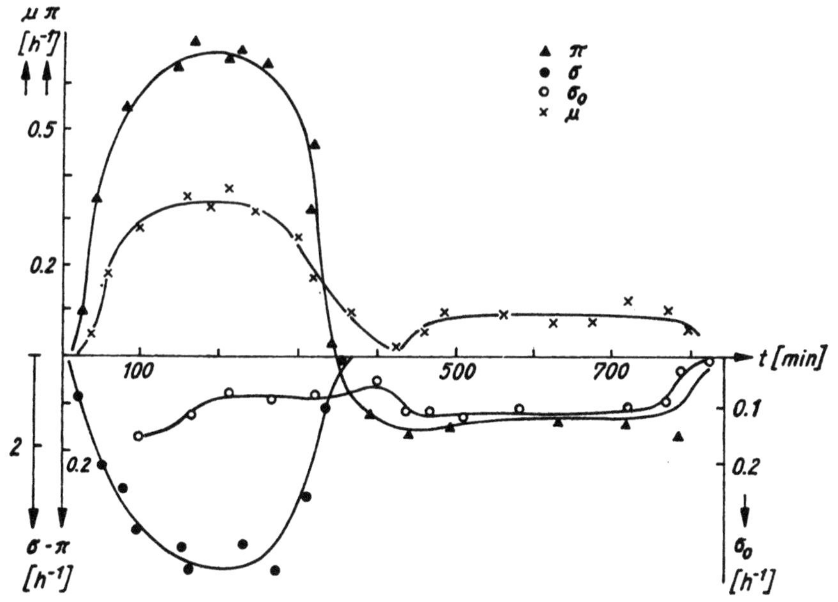

Fig. 6. Quantification of the bioprocess from Fig. 4 (diauxic growth in batch culture) with the aid of the time dependence of the specific rates of growth μ, glucose consumption σ, O_2-consumption σ_O and ethanol production π

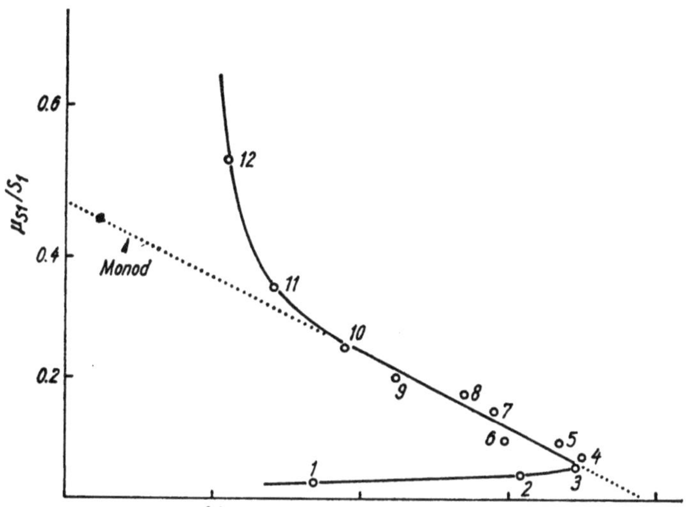

Fig. 7. Identification of the kinetic model for the first diauxic growth phase of the aerobic batch culture of yeast acc. to Fig. 4 from the quantification data on Fig 6: Linearisation of the data in a plot of a Eadie-Hofstee diagram for the basic *Monod* kinetics together with some deviations due to an adaptational lag (points 1,2) and due to simultaneous consumption of glucose and ethanol (11, 12)

the beginning (point 1 and 2, due to a lag) which can be neglected after a period of adaption and at the end (point 11 und 12) caused by simultaneous utilization of ethanol and glucose [27, 28, 29, 30]. This latter effect was shown to be quantified with the aid of a term f_1 using the concept of a critical glucose concentration $S_{1,\mathrm{crit}}$, beyond which value the simultaneous utilization occurs. Therefore, the first growth phase can be quantified by

$$\mu = \mu_{S1} + \mu_{S2} = \mu_{\mathrm{max},1} \frac{s_1}{K_{S1} + S_1} + \mu_{S2} \cdot f_1 \quad (23\mathrm{a})$$

with

$$f_1 = \frac{1}{1 + s_1/s_{1,\mathrm{crit}}}. \quad (23\mathrm{b})$$

The concept of a critical glucose concentration was recently examined in more detail, showing that $S_{1,\mathrm{crit}}$ is strongly depending on $S_{0,1}$, phosphate and nitrogen content of the medium and growth rate μ_{S1} [62, 63]. In this region, therefore, the growth of *Sacch. cerevisiae* is not diauxic.

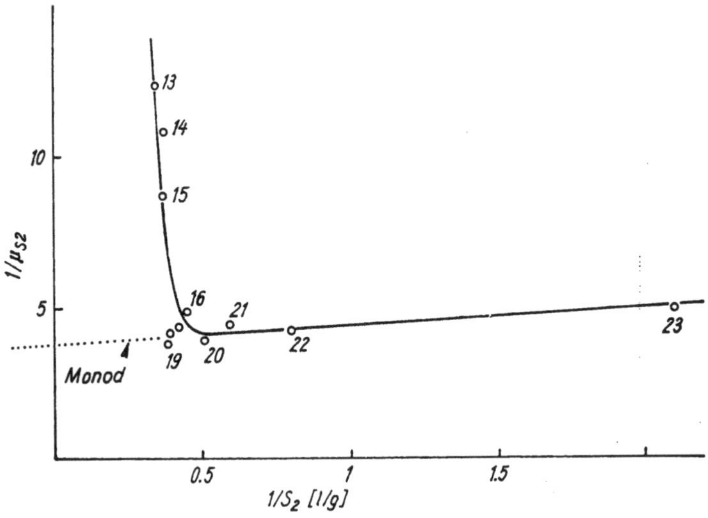

Fig. 8. Model identification and parameter estimation for the second diauxic growth phase of the aerobic batch culture of yeast with ethanol as substrate: Linearisation of the data in a plot of a LINEWEAVER-BURK diagram for the MONOD kinetics with a deviation due to catabolite repression (Points 13—15)

A kinetic analysis of the second growth phase, starting again with the MONOD equation as a first hypothesis, results in a diagram given on Fig. 8. The utilization of ethanol in the second growth phase follows MONOD kinetics but with some deviation in the beginning (point 1—3). The reason for this cannot be found by any kinetic analysis. In case of interest in a mechanistic interpretation this deviation can be shown to be caused by repression of enzymes of the ethanol metabolism (ADH?) due to effects of glucose metabolites. On a formal kinetic level this phenomenon can be quantified arbitrarily with the aid of the following factor f_2, which must be included in eqn. 23a [29]

$$f_2 = \frac{1}{1 + s_2/K_{I2}}. \quad (24)$$

The formal kinetic parameter K_{I2} is taken from an analogy to a substrate inhibition model, which behaves analogous to Fig. 8, but does not represent a "true" inhibition constant and has no mechanistic interpretation. A similar case of the representation of specific product formation rate of penicillin, which is thought to be effected by either catabolite inhibition or repression, with the aid of such a substrate inhibition kinetics is described in literature [4]. However, the equation 23 with 24 is able to fit the data at different conditions as indicated on Fig. 4—5, very well, which fact is illustrated on Fig. 9. Changing the value of the constant K_{I2}, the time dependence of biomass concentration in the diauxic growth can be quantified and adapted to industrial situations.

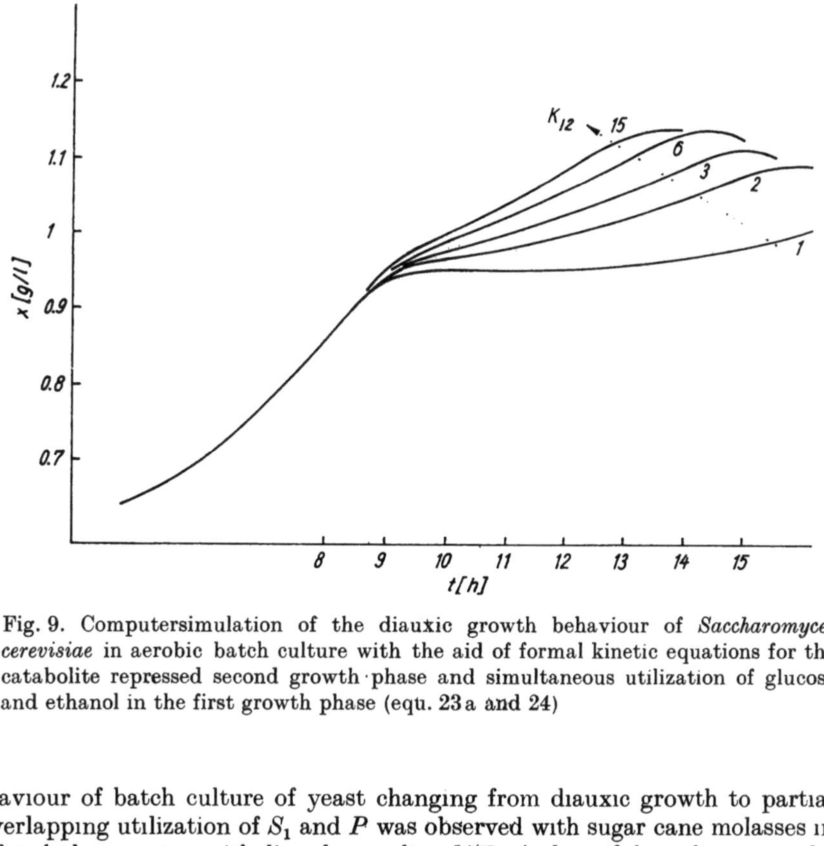

Fig. 9. Computersimulation of the diauxic growth behaviour of *Saccharomyces cerevisiae* in aerobic batch culture with the aid of formal kinetic equations for the catabolite repressed second growth phase and simultaneous utilization of glucose and ethanol in the first growth phase (equ. 23a and 24)

Similar behaviour of batch culture of yeast changing from diauxic growth to partial and total overlapping utilization of S_1 and P was observed with sugar cane molasses in the agitated tubular reactor with liquid recycling [47]. A diauxil lag phase, can be described by arbitrary formal kinetic model functions.
According to Bergter and Knorre [6] the first lag phase is described by eqn. 25:

$$\mu(s_1, t) = \mu_{S_1} \cdot (1 - e^{-t/t_L}) \tag{25}$$

and the second lag

$$\mu(s_i, t) = \mu(s_1, t) + \mu_{S2} \cdot \mu^* \tag{26}$$

with

$$\frac{d\mu^*}{dt} = \frac{1}{t_{L2}} \left(\frac{k_1}{k_1 + s_1} - \mu^* \right). \tag{27a}$$

Eqn. 27 is an analogon to the mathematical function, where biological inertia is described with the aid of formation of ribosoms R [52]

$$\frac{dR}{dt} = \frac{1}{t_L} (\bar{R} - R) \tag{27b}$$

with $\bar{R}$ = stationary value of R.
Another approach is [20]

$$\mu_{S2} = \mu_{\max,2} \frac{s_2}{K_{S2} + s_2} \cdot \frac{s_2}{s_1 + s_2} \tag{28}$$

which is similar to the equation presented here [29]

$$r_x = \mu(s_1, t)\, x + \mu_{S2} \cdot f_1 \cdot f_2 \cdot x \tag{29}$$

Due to the flexibility of eqn. 29, demonstrated in Fig. 9, this approach is advantageous describing real cases of sequential utilization (e.g. diauxie) and partial overlapping of utilization of glucose and ethanol. A general formulation, which is also valid for the case of simultaneous utilization of substrates, cannot be given consequently by this approach. Due to the additive nature of μ_{S1} and μ_{S2} the values of $\mu_{\max,1}$ and $\mu_{\max,2}$ are additive too, which, however, is not the case in reality. Therefore, eqn. 29 should be modified in this direction, which seems to be realized in another approach based on some mechanistic description of multi-substrate utilization [57].
Summarizing the formal kinetic approach of the diauxic growth of *Sacch. cerevisiae* in aerobic batch culture the following equations are thought to be useful:

$$r_{S1} = \frac{ds_1}{dt} = -\frac{1}{Y_{x/S_1}} \cdot \mu_{S1} \cdot x - \frac{1}{Y_{X/P}} \cdot \pi \cdot x. \tag{30}$$

$$r_X = \frac{dx}{dt} = \mu_{S1} \cdot x + \mu_{S2} \cdot x \cdot f_1 \cdot f_2 \tag{31}$$

$$r_P \equiv -r_{S2} = \frac{dP}{dt} = Y_{P/X} \cdot \pi \cdot x - \frac{1}{Y_{X/S_2}} \cdot \mu_{S2} \cdot x \cdot f_1 \cdot f_2 \tag{32}$$

$$r_O = \frac{do}{dt} = k_L \cdot a(o^* - o) - \frac{1}{Y^{I}_{x/o}} \cdot \mu_{S1} \cdot x - \frac{1}{Y^{II}_{x/o}} \cdot \mu_{S2} \cdot x \cdot f_1 \cdot f_2 \tag{33}$$

with $Y_{i/j}$ = yield coefficients
μ, σ and π = specific rates
k_La = volumetric oxygen transfer coefficient
o^* = saturation concentration of O_2 in liquid medium

This set of equations is comparable to the model used for computer simulation of diauxic growth in literature [21].
Last not least, an evaluation of the formal kinetic approach, elaborated here for the case of diauxic growth of *Sacch. cer.* in batch culture, can be given in Tab. 2 by a comparison to mechanistic knowledge of the metabolism. It is interesting to note that in aerobic batch culture of *Sacch. cer.*, the aerobic respiration does not occur or cannot be identified by formal kinetic modeling, due to the fact that in this region ($S_1 \leq S_{1,\text{crit}}$) simultaneous utilization of glucose and ethanol occurs. As a consequence the formal kinetic parameters are perturbated and cannot directly be compared to the values of kinetic parameters found e.g. in chemostat culture of yeast with pure aerobic respiration type of metabolism [9, 61].

Tab. 2. Comparison between formalkinetics and type of metabolism of the diauxic growth of *Saccharomyces cerevisiae* in aerobic batch culture with 1% glucose as limiting substrate

Formalkinetic Type:	maximum value of specific rates (h^{-1})					Metabolic Type
	μ	σ_{S1}	σ_{S2}	π	Y_{XS}	
(1) $S_1 \xrightarrow{O} X, P$	0,38	2–5	0	0,7	0,15	aerobic fermentation
—	0,38	0,7	0	0	0,55	aerobic respiration $S_1 \xrightarrow{O} X$
(2) $S_1 + S_2 \xrightarrow{O} X$	0,38 → 0,05	→ 0	0 → 0,1	(= $-\sigma_{S2}$)	0,5	simultaneous utilization of glucose + ethanol ($s_{1,crit}$)
(3) $S_2 \xrightarrow[K_{I_2}]{O} X$	0,05 → 0,18	0	0,1 → 0,2	(= $-\sigma_{S2}$)	> 0,5	catabolite repressed ethanol utilization
(4) $S_2 \xrightarrow{O} X$	0,18	0	0,2	(= $-\sigma_{S2}$)	> 0,5	ethanol utilization

Conclusions

The fact that effective biokinetics are obviously strongly dependent on the reactor system used in experiments is of great importance in bioreactor design. Data from batch culture cannot be used to design bioreactors of the continuous stirred tank type. Due to the "kinetic similarity" known from chemical engineering, however, batch data can be used in the case of designing continuous plug flow reactors [30, 37]. The reason for this is connected to the general statement that interactions between metabolic reaction inside the microbial cells and physical transports in the bioreactor (the "environment") play the central role [35].

Kinetics and stoichiometry should be investigated in different types of reactors in order to achieve better understanding of these interactions and to receive a more secure guideline for bioreactor design.

Concerning the different levels of investigation, it can be concluded that formal kinetic modeling according to the macroscopic principle is a successful approach when applying an integrating point of view for engineering purpose. Formal kinetic models, normally thought to be only descriptive, receive also a potential predictive character when applying the deductive research methodology, illustrated on Fig. 3. Furthermore it becomes clear that formal kinetics include empirical approaches working with "empty" analogies and/or with speculative partially proved mechanisms. Even if there is a gradual transition with increasing knowledge towards the more mechanistically sustained approach, the formal character of these kinetics applying analogies dominates. Another argumentation concerns the classification of kinetic models, normally made in literature (structured/unstructured, segregated/unsegregated, deterministic/probablistic, corpuscular/continuum models). These subdivisions, however, are not effected by the definition of formal kinetics because they represent a useful supplement. Fig. 10 shows a summarizing scheme of this strategy, which includes biokinetics and bioreactors as representatives for microbial metabolism and physical transport phenomena. The most essential fact are the interactions between biology and physics. This fact causes that both effects must be investigated at the same time. This joint analysis in bioprocessing means that the thinking and working with balancing is the prerequisite of this approach. Stoichio-

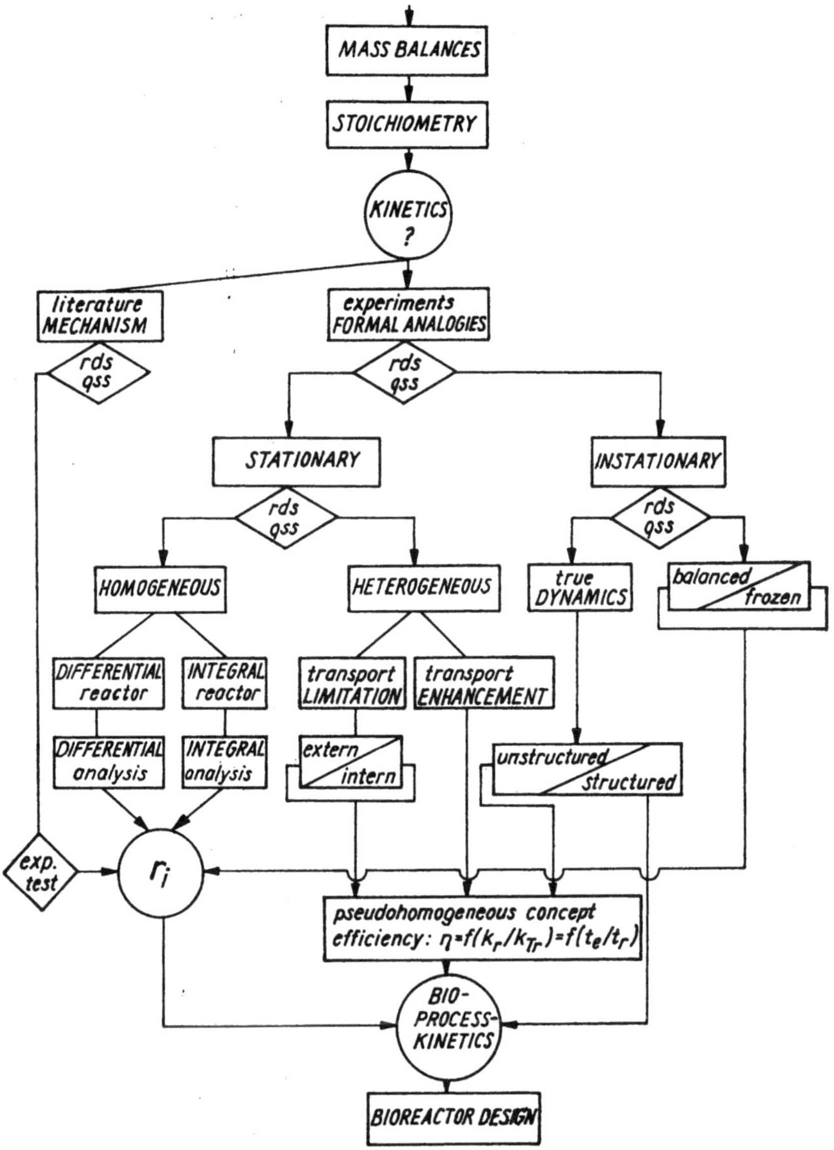

Fig. 10. Flow sheet of the strategy of bioprocess kinetic analysis for different process situations: stationary/instationary, homogeneous/heterogeneous and differential/integral reactor operation

metry on the level of elementary balances or with the overall concept of yield coefficient must be considered consequently together with kinetics. This relationship is manifested by the choice of a certain number of independent rate equations to be sufficient for a full quantification.

The kinetic equations, rarely found in literature, can be derived by postulating mechanisms of reactions or must be extracted from experimental runs. The formal kinetic approach follows the steps of investigations indicated on Fig. 10, where the stationary/

instationary, the homogeneous/heterogeneous and the distributed/lumped parameter (integral/differential reactor resp. reactor with/without gradients) situation must be distinguished. As general principles in the work of elucidating the interactions between metabolism and environment the concepts of rate-determining-step and quasi-steady-state, which are well-known in chemical kinetics, can be successfully applied. The "regime" can be determined where transport limitation or transport enhancement occur in the case of heterogeneous processing or in the case of dynamic mode of operation where "balanced" growth or "frozen" conditions arise. In all these situations a pseudohomogeneous approach is valid, defining efficiencies of the reaction, the transport or the dynamics:

$$\eta = r_{\text{eff}}/r_{\text{ideal}}. \tag{34}$$

The interpretation of these efficiencies in all cases is given by comparing the characteristic values of rate constants of reactions (k_r) or transports (k_{Tr})

$$\eta \propto f(k_r/k_{Tr}) \tag{35a}$$

or the characteristic time of reaction ($t_r \sim 1/k_r$) or environmental change (t_{env})

$$\eta \propto f(t_{\text{env}}/t_r). \tag{35b}$$

This concept was introduced recently [35] and will be outlined, demonstrating the consequences of the formal macroapproach to bioprocessing, in detail in a following paper [36].

Eingegangen: 20. 7. 1982

References

[1] Atkinson, B.: Biochemical reactors, Pion Ltd., London (1974).
[2] Barford, J. P., Hall, R. J.: 7th Australian Cong. Chem. Eng. (1979).
[3] Barford, J. P., Hall, R. J.: Biotech. & Bioeng. **23** (1981) 1735.
[4] Bajpaj, R. K., Reuss, M.: Biotech. & Bioeng. **23** (1981) 717.
[5] Bergter, F.: Wachstum von Mikroorganismen, Jena VEB G. Fischer (1972).
[6] Bergter, F., Knorre, W.: Z. allg. Mikrobiol. **12** (1972) 613.
[7] Eckenfelder, W. W.: Industrial Water Pollution Control, chap 13, Mc Graw-Hill, N. Y., (1966).
[8] Fiechter, A.: Mitt. d. Versuchsanstalt f. d. Gärungsgewerbe **22** (1968) 1.
[9] Fiechter, A.: in Proc. 4th Intern. Symp. Yeasts, Vienna, **II** (1974) 17.
[10] Fiechter, A.: in Handbook of Biotechnology, Vol. 1 (1981) 455 (Rehm, H. J., Reed, G. eds.),
[11] Glansdorff, P., Prigogine, I.: Thermodynamics of structure, stability and fluctuations. Wiley, N. Y. (1974).
[12] Freundlich, H.: in Thomas, J. M.: Introduction to principles of heterogeneous catalysis, Academic, N. Y. (1926).
[13] Harder, A., Roels, J. A.: in Adv. Biochem. Eng. **21** (1981) 56.
[14] Harremoes, P.: Vatten **2** (1977) 122.
[15] Heijnen, J. J., et al.: Biotech. & Bioeng. **21** (1979) 2175.
[16] Heijnen, J. J., Roels, J. A.: Biotech. & Bioeng. **23** (1981) 739.
[17] Hofmann, H.: Chem. Ing. Techn. **47** (1975) 191.
[18] Hofmann, H.: Chem. Ing. Techn. **47** (1975) 823.
[19] Hofmann, H.: Ger. Chem. Eng. **2** (1979) 258.
[20] Imanaka, T. et al.: J. Ferm. Technol. **50** (1972) 633.
[21] Knorre, W.: in Mathematische Modellbildung in Naturwissenschaft und Technik, Akademie Verlag Berlin (1976) 221.
[22] Kossen, N. W. F.: Biotech. Lett. 2, sept. (1980).

[23] Lafferty, R. M., Moser, A., Steiner, W.: in Proc. 4th Intern. Symp, Yeasts, Vienn, **II** (1974) 1.
[24] Monod, J.: Recherches sur la croissance des cultures bacteriennes, Hermann, Paris (1952).
[25] Moreira, A. R. et al.: Biotech. & Bioeng. Symp. **9** (1979) 179.
[26] Moser, A. et al.: in Proc. 4th Intern, Symp, Yeasts, Vienna, **I** (1974) 117.
[27] Moser, A.: Thesis of habilitation, TU Graz (1977).
[28] Moser, A.: Chem. Ing. Techn, **49** (1977) 612.
[29] Moser, A.: in Proc. 1st Europ. Congr. Biotech., Interlaken/Suisse, **I** (1978) 88.
[30] Moser, A.: in Proc. 2nd IWTU Symp., Vol. **2** (1980) 177; (M. Moo-Young et al., eds) Pergamon Press.
[31] Moser, A.: paper 6 th IFS, London/Ont., Canada, (1980).
[32] Moser, F.: Paper at 2nd World Conf. Chem. Eng., 4—8 oct., Montreal, Canada (1981).
[33] Moser, A.: Bioprozeßtechnik, Springer Verlag Wien, New York (1981).
[34] Moser, A.: Biotech. Lett. **4** (1982) 73.
[35] Moser, A.: in Proc. 3rd Chem. Eng. Conf. Austria—Italy. Yugoslavia, Graz, sept. 14—16, (1982).
[36] Moser, A.: Acta Biotechnol. (1984) in press.
[37] Moser, A.: Biotech. Lett. **4** (1982) 282.
[38] Netter, H.: Theoretical Biochemistry, Oliver & Boyd, Edinburgh (1969).
[39] Pamment, N. B. et al.: Biotech. & Bioeng. **20** (1978) 349.
[40] Peringer, P. et al.: Biotech. & Bioeng. Symp. **4** (1973) 27.
[41] Peringer, P. et al.: Biotech. & Bioeng. **16** (1974) 431.
[42] Peringer, P., Blachere, H. T.: Biotech. & Bioeng. Symp. **9** (1978) 205.
[43] Pirt, S. J.: Principles of microbe and cell cultivation, Blackwell, Oxford, (1975).
[44] Popper, K.: Die Logik der Forschung, Mohr Studienausgabe (1934).
[45] Popper, K.: The growth of scientific knowledge, Routledge & Kegan Paul, London (1972).
[46] Ringpfeil, M.: Chem. Techn. **29** (1977) 423.
[47] Rybnicek, R.: M. sc. thesis TU Graz (1979).
[48] Reuss, M.: in Fortschr. d. Verfahrenstechnik **15** F (1977) 549.
[49] Roels, J. A.: Biotech. & Bioeng **22** (1980) 2457.
[50] Roels, J. A., Kossen, N. W. F.: in Progr. Ind. Microb. **14** (1978) 95.
[51] Roels, J. A.: J. Chem. Tech. Biotechnol. **32** (1982) 59.
[52] Romanovsky, J. et al.: Kinetische Modelle in der Biophysik, VEB G. Fischer, Jena (1974).
[53] Sylvester, N. D. et al.: Can. J. Chem. Eng. **53** (1975) 313.
[54] Theophilou, J., Wolfbauer, O. Moser, F.: Gas-Wasser-Fach, Wasser/Abwasser (1978) 119, 135.
[55] Toda, K. et al.: Biotech. & Bioeng. **22** (1980) 1805.
[56] Toda, K.: J. Chem. Tech. Biotechnol. **31** (1981) 775.
[57] Tsao, G. T., Hanson, Th.: Biotech. & Bioeng. **17** (1975) 1591.
[58] van Dedem, G., Moo-Young, M.: Biotech. & Bioeng. **17** (1975) 1301.
[59] van Krevelen, D. W.: in Proc. 1st Symp. Chem. Reaction Eng., A'dam, Pergamon Press, (1957).
[60] von Bertalanffy, L.: Biophysik des Fließgleichgewichtes, Vieweg (1953).
[61] von Meyenburg, K.: Dissertation 4279, ETH Zürich (1969).
[62] Wöhrer, W., Röhr, M.: Proc. 4th FEMS Symp., Vienna B2 (1977).
[63] Wöhrer, W., Röhr, M.: Biotech. & Bioeng. **23** (1981) 567.
[64] Wolfbauer, O. et al.: Chem. Eng. Sci. **33** (1978) 953.
[65] Wuhrmann, K., et al.: Schweiz. Z. Hydrol. **20** (1958) 284.

Acta Biotechnol. 3 (1983) 3, 217–220

Activity of Diaminopimelic Acid Decarboxylase in Bacteria Producing Lysine

S. K. Sen, M. Chatterjee, S. P. Chatterjee and A. K. Banerjee

Microbiology Laboratory, Department of Botany, Burdwan University, Burdwan 713104, India

Summary

Activity of DAP-decarboxylase in lysine overproducing strains of *Micrococcus luteus, M. varians* and *Arthrobacter globiformis* Lb reached the highest level during the end of the exponential phase of growth and remained at the same high level during the stationary phase of growth when major bulk of lysine was accumulated. In comparison in a lysine non-producing strain of *Arthrobacter globiformis* I_4 the activity of the same enzyme was low. DAP-decarboxylase of these three lysine overproducers has two specialities, persistence during the stationary phase and insensitivity to repression by exogenous lysine.

Zusammenfassung

Die Aktivität der Diaminopimelinsäure-Decarboxylase wurde bei den Lysin-überproduzierenden Stämmen *Micrococcus luteus, M. varians* und *Arthrobacter globiformis* Lb. untersucht.
Maximalwerte wurden am Ende der exponentiellen Wachstumsphase erreicht; während der stationären Wachstumsphase blieben diese Werte erhalten. Die Hauptmenge an Lysin wurde in der stationären Phase ausgeschüttet.
Bei einem nicht Lysin produzierendem Stamm (*Arthrobacter globiformis* I_4) blieb die untersuchte Enzymaktivität niedrig. Die Diaminopimelinsäure-Decarboxylase der drei überproduzierenden Stämme ist durch folgende Charakteristika ausgezeichnet:

- Erhaltung der hohen Aktivität in der stationären Wachstumsphase
- Unempfindlichkeit gegenüber hohen extracellulären Lysin-Konzentrationen (keine feed-back-Hemmung).

Introduction

The last enzyme of the lysine biosynthetic pathway, diaminopimelate decarboxylase (meso-2,6 diaminopimelate carboxylase, EC 4.1.1.20) specifically decarboxylates meso isomer of diaminopimelic acid (DAP) to yield lysine [13]. In most organisms the activity of this enzyme is highest at the end of the exponential phase of growth and thereafter it rapidly declined [4, 10]. The activity of the enzyme is higher in microorganisms grown in defined minimal media than in those grown in nutrient medium [1, 10]. The synthesis of DAP-decarboxylase has been shown to be repressed by lysine in *Escherichia coli* [8, 10]; *Aerobacter aerogens* [10]; *Streptococcus faecalis* [3]; most of the species of *Bacillus*

[4, 5] and in the lysine-overproducing homoserine auxotroph of *Micrococcus glutamicus* [2].

Present communication reports the activity of DAP-decarboxylase under different stages of growth of three strains of hydrocarbon utilizing lysine-excreting microorganisms: *Micrococcus luteus* B_6, *M. varians* 2Fa, and *Arthrobacter globiformis* Lb. Influence of exogenous lysine on the DAP-decarboxylase activity of these microorganisms has been studied and compared with that of a lysine non-producing isolate of *A. globiformis* I_4.

Materials and Methods

Organisms, maintenance and cultivation

Lysine overproducing microorganisms were isolated from soil by bioautographic technique [9] using a lysine auxotroph of *E. coli* K_{12}. *A. globiformis* I_4, a glutamic acid excretor, was isolated similarly using a glutamic acid auxotroph of *E. coli* K_{12}. The organisms were maintained at 30 °C on agar slants of the medium of MAKULA and FINNERTY [7] with 1% (w/v) glucose as C-source.

Cultures were grown in 1 l flasks with 200 ml liquid medium [7] containing glucose (1% w/v) and supplemented with biotin (1 $\mu g\ l^{-1}$). The flasks were inoculated with an aqueous suspension of bacteria from 24 h old slant culture and incubated at 30 °C on a rotary shaker.

Growth and lysine assay

Growth was measured in terms of dry weight of washed bacterial cells. Extracellular lysine was measured colorimetrically [12].

Assay of DAP-decarboxylase

Cells were harvested by centrifugation at 10,000 · g for 60 min and washed twice with 0.2 M sodium-phosphate buffer pH 7.3. The cell-free extracts were prepared from cells frozen overnight by crushing them in sand with half of its volume of 0.2 M phosphate buffer and extract was centrifuged at 10,000 · g for 20 min at 2 °C. Protein in the extract was measured by the method of LOWRY et al [6]. DAP-decarboxylase activity was meassured by the colorimetric method of WHITE [11].

Results and Discussion

DAP-decarboxylase of the lysine overproducing microorganisms increased during the growth. The activity showed a sharp rise (Table 1) during 12 to 24 h and reached a maximum at the end of the exponential phase of growth. In all the cases extracellular lysine began to accumulate at the end of the exponential phase. The major bulk of lysine accumulated in the medium during the period between 36 to 48 h. After the end of the exponential phase of growth DAP-decarboxylase activity remained almost at the same level for *M. varians*, but declines to about 75% of its maximal activity in the case of *M. luteus* and *A. globiformis* Lb. Such a steady level of DAP-decarboxylase activity during the stationary phase of growth has been observed in *Staphylococcus aureus* [6] and lysine producing homoserine auxotroph of *M. glutamicus* [2].

Table 1. DAP-decarboxylase activity during growth and lysine production by lysine over producing microorganisms

Microorganisms	Time, h						
	0	12	24	36	48	60	72
M. luteus B_6							
Growth[a]	1	11	46	90	110	111	110
DAP-decarboxylase activity[b]	10	25	75	110	100	90	80
Extracellular lysine[c]	—	—	—	15	56	60	60
M. varians 2 Fa							
Growth	0.8	25	38	43	45	45	44
DAP-decarboxylase activity	1	15	20	27	30	30	30
Extracellular lysine	—	—	—	18	32	36	38
A. globiformis Lb							
Growth	0.9	20	42	48	52	53	53
DAP-decarboxylase activity	1	15	30	40	52	40	40
Extracellular lysine	—	—	—	20	46	48	48

[a] Dry weight, mg/100 ml; [b] Specific activity, nM of DAP-decarboxylated mg of protein/min; [c] mg/100 ml.

Comparison of DAP-decarboxylase activity of the lysine excreting strains with that of a lysine non-excreting *A. globiformis* I_4, grown with different levels of lysine revealed (Table 2) that in DAP-decarboxylase activity of the former group was distinctly higher than the non-producing strain. Moreover, it was also seen that the DAP-decarboxylase of the non-producing strain was repressed by exogenous lysine, whereas the enzymes of the lysine excreting strains were insensitive to lysine repression. Lysine insensitive DAP-decarboxylases have been reported [10] in *B. megaterium* and *B. subtilis*, though they are not lysine overproducers.

It may be concluded that DAP-decarboxylase of the lysine overproducing strains has two important properties: first, it is insensitive to lysine repression, and second it persists during the stationary phase of growth. The decline of DAP-decarboxylase activity in *M. luteus* B_6 and *A. globiformis* Lb after the exponential phase of growth was not due to the regulatory effect of lysine accumulated in the medium, because different concentrations of lysine added to the medium did not repress the enzyme activity. Most likely, this decline of the enzyme activity is due to exhaustion of cell metabolites, due

Table 2. Effect of lysine added in the growth medium on DAP-decarboxylase activity

Lysine added in growth medium[a]	m-DAP-decarboxylase activity[b]			
	M. varians 2 Fa	*M. luteus* B_6	*A. globiformis* Lb.	*A. globiformis* I_4
0	36	118	45	20
10	38	120	60	20
20	40	120	62	20
50	40	120	60	15
100	40	121	60	10
200	40	121	60	2

[a] μg/ml; [b] nM of DAP-decarboxylated/mg of protein/min, at 48 h.

to normal ageing of cells, which include diminished activity at an early steps of lysine biosynthetic pathway. For the cessation of DAP-decarboxylase activity in lysine excreting homoserine auxotroph of *M. glutamicus*, a similar explanation has been put forward [2]. Since lysine does not at all repress the DAP-decarboxylation step, it may also be concluded that in these lysine excretors the main site(s) of regulation of lysine biosynthesis is located at the earlier steps of pathway.

Acknowledgements

The authors are grateful to the UGC and CSIR, India, for their financial assistance to M. C. and S. K. S. for their investigation. We are also thankful to Mr. D. K. ROY for supplying the culture of *A. globiformis* I_4.

Eingegangen: 23. 7. 1982

References

[1] DEWEY, D. L., HOARE, D. S., WORK, E.: DAP-decarboxylase in cells and extracts of *Escherichia coli* and *Aerobacter aerogens*; Biochem. J. **58** (1954) 523.
[2] DRAZIC, M., VITALE, L.: The role of diaminopimelate decarboxylase in the regulation of lysine biosynthesis in *Micrococcus glutamicus*; Period. Biol. **76** (1974) 85.
[3] GILBOE, D. P., FRIEDE, J. D., HENDERSON, L. M.: Effect of hydroxylysine on the biosynthesis of lysine in *Streptococcus faecalis*; J. Bacteriol. **95** (1968) 856.
[4] GRANDGENNET, D. P., STAHLY, D. P.: Diaminopimelic decarboxylase of sporulating bacteria; J. Bacteriol. **96** (1968) 2099.
[5] GRANDGENNET, D. P., STAHLY, D. P.: Repression of diaminopimelate decarboxylase by l-lysine in different *Bacillus* species; J. Bacteriol. **105** (1971) 1211.
[6] LOWRY, O. H., ROSENBROUGH, N. J., FARR, A. L., RANDALL, R. J.: Protein measurement with the Folin phenol reagent; J. Biol. Chem. **193** (1951) 265.
[7] MAKULA, R., FINNERTY, W. R.: Microbial assimilation of hydrocarbons. I Fatty acids derived from n-alkanes; J. Bacteriol. **95** (1968) 2108.
[8] PATTE, J. C., LOVINY, T., COHEN, G. N.: Répression de la décarboxylase de l'acide méso-α, ε-diaminopimélique par la l-lysine, chez *Escherichia coli*; Biochem. Biophys. Acta. **58** (1962) 359.
[9] UDAKA, S.: Screening methods for microorganisms accumulating metabolites and its use in isolation of *Micrococcus glutamicus*; J. Bacteriol. **79** (1960) 751.
[10] WHITE, P. J., KELLEY, B., SUFFLING, A., WORK, E.: Variation of the activity of bacterial diaminopimelate decarboxylase under different conditions of growth; Biochem. J. **91** (1964) 600.
[11] WHITE, P. J.: Diaminopimelate decarboxylase (*Escherichia coli*). Methods in Enzymology. Ed. COLOWICK, S. P. and KAPLAN, N. O., New York, Academic Press, 1971. Vol. 17b. p. 140.
[12] WORK, E.: Reaction of ninhydrin in acid solutions with two amino groups and its application to the estimation of α-ε-diaminopimelic acid; Biochem. J. **67** (1957) 416.
[13] WORK, E.: Diaminopimelate decarboxylase. Methods in Enzymology. Ed. by COLOWICK, S. P. and KAPLAN, N. O., New York, Academic Press, 1962. Vol. 5. p. 864.

Acta Biotechnol. 3 (1983) 3, 221–225

Production of Neutral and Alkaline Proteases in Batch and Chemostat Cultures by Bacillus mesentericus 76

E. Emanuilova and A. Kaimaktchiev

Bulgarian Academy of Sciences, Institute of Microbiology
1113 Sofia, Acad. G. Bontchev str. bl. 26

Summary

The kinetics of the bacterial extracellular protease synthesis (neutral and alkaline protease of *Bacillus mesentericus* strain 76, R-form) in batch and chemostat cultures under conditions of glucose limitation were investigated. When the medium was supplemented with casein the production of the proteases was significantly higher. Optimal dilution rates for obtaining of two proteases are fixed. The synthesis of both alkaline and neutral proteases is controlled by catabolite repression and induction.

Zusammenfassung

Die Biosynthese von neutralen und alkalischen Proteasen unter Glucose-Limitation wurde in diskontinuierlicher und kontinuierlicher Kultivierung untersucht. Bei Zusatz von Casein zum Medium wurde eine deutlich höhere Protease-Produktion beobachtet. Für die Bildung von zwei Proteasen wurden optimale Durchflußraten ermittelt. Sowohl die alkalische als auch die neutrale Protease werden durch Katabolitenrepression und Induktion reguliert.

Introduction

The kinetics of bacterial extracellular protease synthesis in batch cultures were investigated to a great extent [1]. The limited number of studies on this synthesis in continuous cultures may be explained to a certain extent by the fact that in chemostat cultures the enzyme production is decreasing due to the influence of many factors. For achieving optimal productivity of extracellular microbial enzymes in chemostat cultures of great importance are: the type and concentration of limiting substrate, the influence of dilution rate on the physiological changes of the culture, the formation of intracellular inducers and repressors, the rate of synthesis of other intra- and extracellular metabolites [2].

The main purpose of this study is to clear up some aspects of the regulatory processes on the synthesis of neutral and alkaline proteases in batch and chemostat cultures by *B. mesentericus* in conditions of glucose limitation.

Materials and methods

MICROORGANISM. *Bacillus mesentericus* (According to "Bergey's Manual of Determinative Bacteriology" R. E. Buchanan and Williams, 1974, USA, Baltimore, the species *mesentericus* is regarded as a synonym of *subtilis*.) strain 76,R-form, producing neutral and alkaline proteases was used.

MEDIUM. The cultivation medium used had composition (g/l) as follows: $(NH_4)_2SO_4$-1.0; K_2HPO_4-5.65; KH_2PO_4-3.38; $MgSO_4$-0.8; glucose-8.0; $CaCl_2$-0.15; Fe_2SO_4-0.001; $ZnSO_4$-0.01; $MnCl_2$-0.05. In order to study the influence of casein on protease synthesis, the medium was supplemented with 5.0 Difco casein, turned to a soluble Na-salt.

CULTIVATION. For inoculum 1% v/v of 16—18 hours culture of the strain, cultivated on a gyratory shaker at 240 rev/min in an Erlenmeyer flask at 37 °C was used. Batch and continuous cultures were carried out in an Ankum-2 fermentor of 1500 ml working volume with sterile air flow kept at 0.2 v/v. The impeller speed was maintained at 600 rev/min and the temperature was 37 °C. The turn to continuous culture was carried out on the 11th hour of inoculation in the case of synthetic medium and on the 5th hour in the case of medium supplemented with casein. Bacteria were cultured at each dilution rate for at least three volume changes. After achieving "steady state" 3 parallel samples were collected at hourly intervals for three hours.

ANALYTICAL PROCEDURES. The sample of the culture broth was centrifugated and supernatant fluid was used for determination of enzyme activity. The biomass tractions were dried at 105 °C to constant weight. Proteolytic activity was assayed by the method of Hagihara and al. [3], and is expressed in PUN · ml^{-1}. The activity of neutral and alkaline proteases was measured by treatment with EDTA and inhibitor isolated from potatoes by the method of Natchev and al. [4]. Glucose was assayed by the method of Somogyi for reducing sugars [5].

Results and discussion

Figure 1a and 1b shows the results obtained from batch culture. In synthetic medium (Fig. 1a) the production of proteases starts when the culture enters exponential phase of growth. The highest rate of enzyme production was achieved at the end of this phase and at the beginning of the stationary one. This relation was investigated for this strain in our previous studies [6] and this is in accordance with the results of other authors with *B. subtilis* [1]. The alkaline protease had a lower level of activity and its synthesis was stopped earlier than the neutral one's.

With the addition of casein in the medium (Fig. 1b) a reducing of the lag phase of the culture was observed and the production of the neutral and alkaline proteases was significantly higher. The synthesis started earlier when the glucose concentration in the medium was higher. The effect of catabolite repression in the presence of casein probably is not so strongly expressed in comparison with the cultures without casein. Moreover the synthesis of both proteases proceeded longer.

The results obtained from chemostat cultures are presented on figures 2a and 2b. The figures show the effect of dilution rate on the specific rate of enzyme synthesis. The highest rate of production of both proteases in the medium with casein is at dilution rate 0.4 h^{-1}. In medium without casein this maximum is the same for the neutral protease and at 0.35 h^{-1} for the alkaline one.

The production of both proteases is lower in continuous culture, but the rates of their synthesis on both types media are higher, compared with those, obtained in batch culture. With decreasing dilution rates the rates of synthesis of both proteases decrease too.

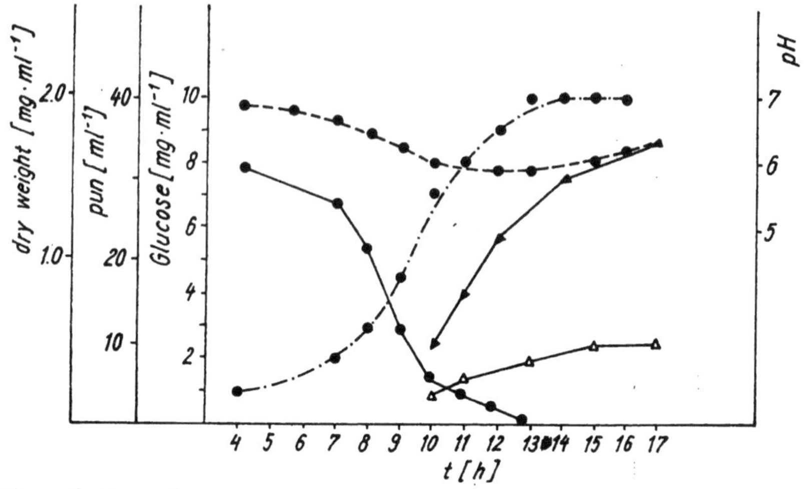

a) synthetic medium

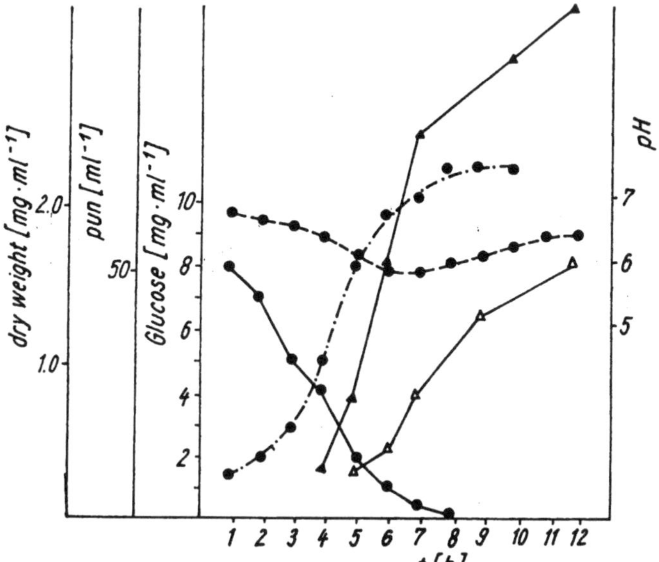

b) medium with casein

Fig. 1. Production of biomass and proteases in batch culture: biomass-dry weight, (1) neutral protease, (2) alkaline protease, (3) pH of the culture, (4) glucose concentration (5) in culture supernatant

(1) ●···●; (2) ▲——▲; (3) △——△; (4) ●--●; (5) ●——●

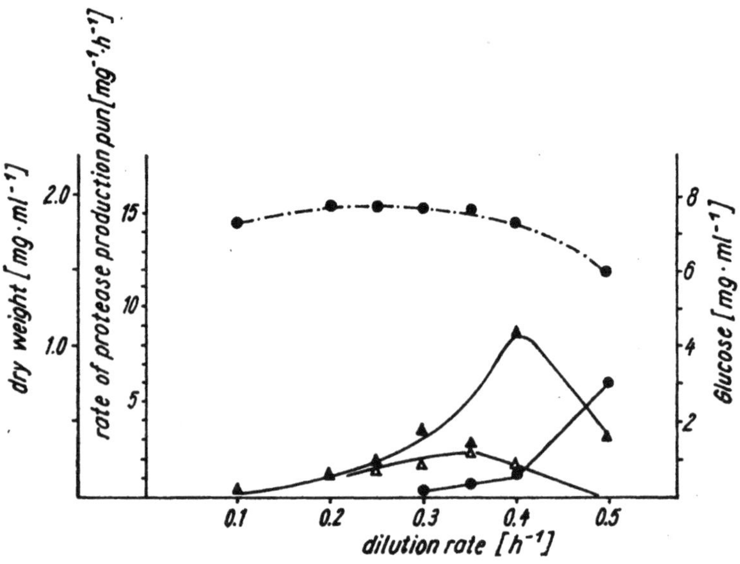

a) synthetic medium

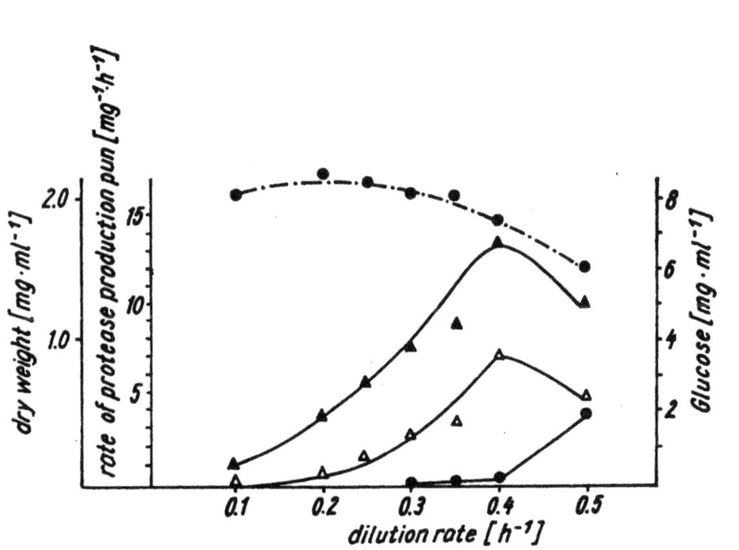

b) medium with casein

Fig. 2. Effect of dilution rate on biomass production and rate of enzyme production: biomass-dry weight, neutral protease, alkaline protease, glucose concentration in culture supernatant

Figure 3 shows the relationship between dilution rate and the rate of enzyme synthesis in continuous culture [7]. Apparently the pattern of the curves presenting their rates of synthesis of the alkaline and neutral proteases in our work corresponds to that of curve D. By the results obtained so far it is evident that the synthesis of both alkaline and neutral proteases of *Bacillus mesentericus* 76 is controlled by catabolite repression and induction.

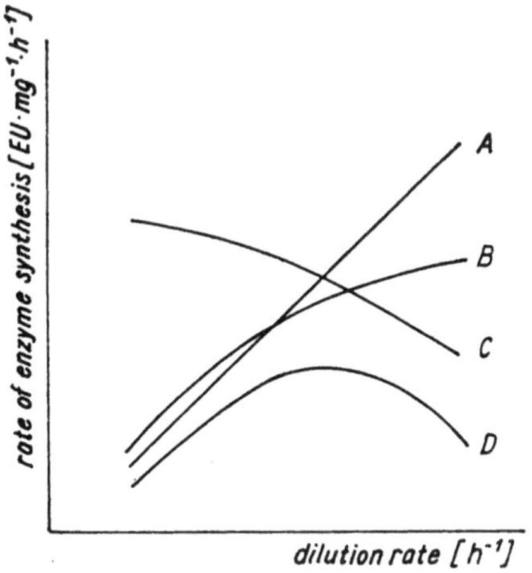

Fig. 3. General patterns of rate of enzyme production in continuous culture
A. Constitutive enzymes whose synthesis only depend on cell concentration.
B. Inducible enzymes.
C. Constitutive enzymes, sensitive of catabolite repression.
D. Inducible enzymes, sensitive to catabolite repression.

Eingegangen: 30. 4. 1982

References

[1] Priest, F. G.: Bacteriol. Rev. **3** (1977) 711.
[2] Логинова, Л.: Теория и практика непрерывного культивирования микроорганизмов. Москва (1980) 126.
[3] Hagihara, B., Matsubara, H., Nakavi, H., Okunuki, K.: J. Bacteriol. **3** (1958) 185.
[4] Начев, Л., Велчева, П., Даов, Т., Костуркова, П., Колев, Д.: Сб. III конгрес по микробиология (1975) 101.
[5] Somogyi, M.: J. Biol. Chem. **195** (1952) 19.
[6] Емануилова, Е., Каймакчиев, А.: Acta microbiol. bul. **8** (1981) 51.
[7] Wiersma, W.: Ph. D. thesis. Univ. of Gröningen (1978) 29.

Acta Biotechnol. 3 (1983) 3, 226

The Leipzig Symposium on Biotechnology 1982

Sept. 26–29th, Leipzig, G.D.R.

The utilization of hydrocarbons by microorganisms, which use them as the sole source of carbon and energy, has been well known since the beginning of our century. Originally it was widely believed that this phenomenon was the peculiar ability of only a few specialized microorganisms.
Today, however, it is established knowledge, that hydrocarbon utilization occurs in nearly all taxonomic groups of free-living bacteria, yeasts and fungi.
The use of these abilities, mainly in the industrial production of SCP and of other valuable products such as organic acids and tensides and their application to the removal of environmental pollutants caused a considerable heightening of interest in these so-called "unconventional substrates". Moreover, this could be recognized by the worldwide increase of basic research in this field, especially in microbiology, biochemistry and genetics. The Leipzig Symposium on Biotechnology 1982, held from September 21st to 25th, was dedicated to the topic of "Microbiology and Biochemistry of Hydrocarbon Conversion". It was the aim of the symposium to elucidate the present situation and to bring about an exchange of ideas concerning the prospects of such substrates as n-alcanes, gas-oil, methane and methanol in biotechnology.
The organizers of the Leipzig Symposia on Biotechnology are the Society of General and Industrial Microbiology in the Biological Society of the G.D.R (Special Group of Biotechnology), the Institute of Technical Chemistry of the G.D.R Academy of Sciences and the Department of Biological Sciences of the Karl Marx University of Leipzig.
According to the intention of the organizers the symposia shall be held every two years. In any case there should be a forum of biotechnology in the GDR to foster the publication of recent findings as well as further education and training.
In 1984 the topic of this symposium will be "Phages in Industrial Microbiology".
This year participants from GDR institutes of the universities, the Academy of Sciences and industrial laboratories together with invited scientists from abroad, including the Soviet Union, France, Japan, Great Britain, the FRG and Switzerland, presented 26 lectures and 70 posters.
In the final round table discussion the participants come to the conclusion that hydrocarbons and methanol will continue be important substrates, which will surely exert an essential influence upon the further development of biotechnology.
In the present and the following volumes of "Acta Biotechnologica" some selected lectures of the Leipzig Symposium on Biotechnology 1982 will be published.

H.-P. Kleber
K. Sattler

Acta Biotechnol. 3 (1983) 3, 227–240

SCP-Produktion auf der Basis von Kohlenwasserstoffen

M. Ringpfeil

Akademie der Wissenschaften der DDR
Insitut für technische Chemie Leipzig
7050 Leipzig, Permoserstr. 15

Biotechnologie-Symposium 1982, Leipzig, DDR

Summary

Microbial synthesis of SCP on hydrocarbons is regaining importance. Improved methods of energetic balancing allow its economic assessment also under conditions denying direct proportionality between costs and social expenditures. Main methods for augmenting economy of SCP synthesis are recovery of further products beyond SCP, simplification of downstream processes, and rise in yield and productivity of fermentation.

Zusammenfassung

Die mikrobielle Eiweißsynthese auf Kohlenwasserstoffen gewinnt wieder an Bedeutung. Verbesserte Methoden der energetischen Bilanzierung erlauben ihre ökonomische Einschätzung auch unter Bedingungen, unter denen die direkte Proportionalität zwischen Kosten und gesellschaftlichen Aufwendungen nicht gegeben ist.
Hauptmethoden zur Erhöhung der Ökonomie der mikrobiellen Eiweißsynthese sind die Gewinnung von mehr Produkten als nur Eiweiß, die Vereinfachung der Aufarbeitungsprozesse und die Steigerung von Ausbeute und Produktivität der Fermentation.

Резюме

В настоящее время микробный синтез белка на углеводородах опять приобретает значение. Усовершенствованные методы энергетического балансирования позволяют его экономи ческую оценку даже в условиях, в которых не имеется прямой пропорциональности между стоимостью и общественными затратами.
Главными методами по повышению экономической эффективности микробного синтеза белка являются получение дополнительных продуктов наряду с белком, упрощение последующих процессов, повышение выхода и продуктивности ферментации.

Die Stellung des mikrobiellen Eiweißes[1] gegenüber pflanzlichen und tierischen Alternativprodukten

Die industrielle Erzeugung von mikrobiellem Eiweiß aus fossilen Rohstoffen ist der erste Versuch, Nahrungsmittel in größerem Umfang ohne Inanspruchnahme landwirtschaftlicher Methoden oder Vorprodukte zu erzeugen. Nach dem aufsehenerregenden Beginn dieser Entwicklung in den 60er und frühen 70er Jahren ist Mitte der 70er Jahre ein ebenso spektakulärer Abbruch erfolgt. Nur in der UdSSR ist eine systematische und rasche Weiterentwicklung erfolgt. 1980 betrug dort die Produktion bereits mehr als 1 Mio t [1]. Für den Abbruch dieser Entwicklung in den USA, Japan und Westeuropa werden politische, ökonomische und gesundheitliche Gründe angeführt, deren Gültigkeit umstritten ist [2]. Das Ansteigen der Erdöl- und Erdgaspreise im Zusammenhang mit der Erdölkrise der 70er Jahre hat zweifellos Auswirkungen gehabt [3]. Die Rohstoffkosten für die mikrobielle Eiweißproduktion liegen heute in der Größenordnung der Kosten für vergleichbare landwirtschaftlich erzeugte Produkte (Tab. 1) [4]. Der Spielraum für die Gestaltung einer ökonomischen industriellen Produktion ist geringer geworden. Er ist aber — je nach dem Rohstoffaufkommen und der Rohstoffverfügbarkeit der einzelnen Länder — kleiner oder größer vorhanden und erlaubt die Inangriffnahme der Produktionsentwicklung. Besonders günstige Bedingungen werden in den Erdölförderungsländern gefunden, wo z. B. die begleitenden Erdgase heute noch überwiegend einen ungenutzten Abfall bei der Erdölförderung darstellen [5].

Tabelle 1. Preise für Eiweißfuttermittel und für fossile Rohstoffe zur mikrobiellen Eiweißsynthese (Bezugsjahr 1981) in \$/t

Fischmehl	450
Sojaschrot	225
Erdgas/Methan	220
Propan	250
Normalbutan	320
Dieselöl	360
n-Paraffine	380 (1978)
Methanol	245
Ethanol	725

Mikrobielles Eiweiß ist als ein aufkommendes Alternativprodukt zu eingeführten pflanzlichen und tierischen Eiweißen zu betrachten. Als wichtigstes Vergleichsprodukt aus dem landwirtschaftlichen Bereich gilt Sojaschrot oder -mehl, das ein Nebenprodukt ker Sojaölgewinnung ist. Es besitzt heute den weitaus größten Anteil an der Eiweißdonzentratproduktion (Tab. 2) [6]. Die USA und Brasilien sind die absoluten Haupt-

Tabelle 2. Eiweißkonzentratproduktion 1981/82 (44% Protein) in Mio t

Sojabohnen	60,2
Baumwollsamen	8,1
Erdnüsse	4,3
Sonnenblumenkerne	4,6
Raps	4,6
Fisch	6,8

1 Hierunter wird in diesem Beitrag mikrobielle Biomasse mit einem Rohproteingehalt von minmindestens 44% verstanden.

erzeuger (Tab. 3) [6]. Es ist gerechtfertigt, von einem Sojamonopol und sogar von einem Eiweißkonzentratmonopol zu sprechen. Die Einführung des mikrobiellen Eiweißes wird dadurch erschwert.

Tabelle 3. Sojabohnenproduktion und -verarbeitung 1981/82 in Mio t

USA	Ernte	55,3
Brasilien	Ernte	18,5
Gesamt	Ernte	86,8

Als Rohstoffe für die mikrobielle Eiweißsynthese aus der Gruppe der Kohlenwasserstoffe kommen heute Methan [7], Propan/Butan [8], Pentan/Hexan/Heptan [9], Dieselkraftstoff [10] und gereinigte längerkettige n-Paraffine [11] in Betracht. Mit zunehmender Verarbeitung steigen die Kosten für diese Rohstoffe (s. auch Tab. 1) (Abb. 1) [12]. Bis jetzt haben jedoch nur Verfahren auf der Basis von Dieselkraftstoff und gereinigten n-Paraffinen technische Dimensionen erreicht, darunter das Fermosin®-Verfahren der

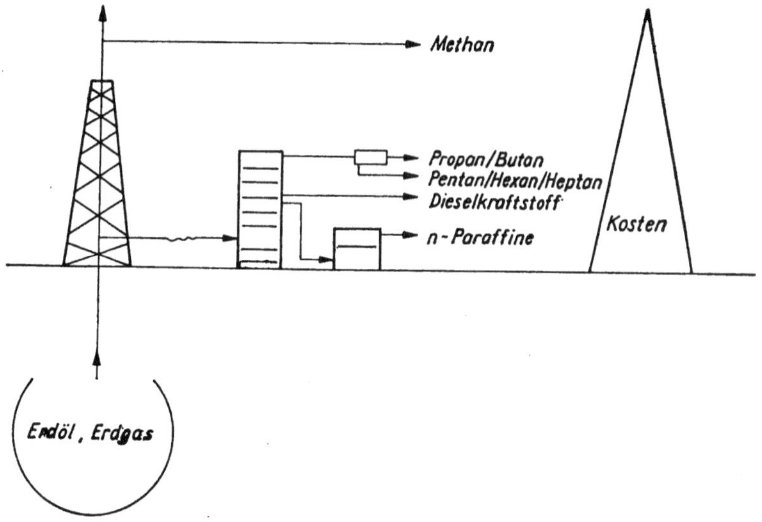

Abb. 1. Kohlenwasserstoffe für die mikrobielle Eiweißsynthese

DDR/UdSSR zur Gewinnung von Futterhefe aus Dieselkraftstoff [13]. Sie werden ausschließlich mit Hefen durchgeführt. Die Konvertierung der leichten Kohlenwasserstoffe erfolgt mit Bakterien. Die Verfahren auf der Basis von Erdgas (Methan), die von einer Gruppe von Instituten und Betrieben der UdSSR und DDR ausgearbeitet wurden, haben technische Reife erreicht.

Zur Ökonomie der mikrobiellen Eiweißproduktion

Die mikrobiellen Eiweißprodukte können und werden sich gegenüber den landwirtschaftlichen nur dann durchsetzen, wenn sie geringere gesellschaftliche Aufwendungen erfordern. Diese können in einige Hauptfaktoren untergliedert und in einer Ungleichung zusammengefaßt werden.

$$A_{\substack{\text{landwirtschaftlich}\\ \text{erzeugte Futtermittel}}} > A_{\text{Kohlenstoffquelle}} + A_{\substack{\text{Betrieb}\\ \text{Invest}\\ \text{Erhaltung}}} - E_{\text{andere Produkte}}$$

Sowohl die Aufwendungen für das (Vergleichs-) Futtermittel aus der landwirtschaftlichen Produktion als auch für die Kohlenstoffquelle zur mikrobiellen Eiweißproduktion[2] stellen für den Produzenten mikrobiellen Eiweißes nicht beeinflußbare Größen dar. Aus ihrer Differenz ergibt sich der Spielraum, den der Produzent mikrobiellen Eiweißes hat, um die Ungleichung zu erhalten. Dieser Spielraum kann vergrößert werden, wenn im Zuge der Herstellung mikrobiellen Eiweißes weitere Produkte gewonnen werden, die einen Erlös bringen. Je fortgeschrittener der wissenschaftlich-technische Stand des Verfahrens und der Anlage sind, um so geringer gestalten sich die Aufwendungen für Investition, Betrieb und Erhaltung der mikrobiellen Produktion, um so geringer ist die Wahrscheinlichkeit, daß mehr Kohlenstoffquelle verbraucht wird und um so höher stellt sich der Erlös aus zusätzlichen Produkten.

Die gesellschaftlichen Aufwendungen werden nicht immer und nicht überall durch die aktuellen Kosten ausgedrückt. Die richtige Berechnung oder auch nur Abschätzung der gesellschaftlichen Aufwendungen ist deshalb eine komplizierte Aufgabe. In solchen Fällen wird versucht, auf naturwissenschaftlich und technisch definierte Kennziffern zurückzugehen, deren Proportionalität zum gesellschaftlichen Aufwand offensichtlich ist und auf deren Basis einen Vergleich der zu betrachtenden Prozesse anzustellen [14]. Er beruht auf der Bilanzierung solcher Kennziffer(n) über die betrachteten Prozesse. Für die Maisproduktion wurde z. B. eine Bilanzierung des Energieverbrauchs unter den Bedingungen der amerikanischen Landwirtschaft vorgenommen [15]. Daraus ergibt sich ein Vergleich, der in jedem Fall zuungunsten mikrobieller Eiweißproduktionen aus fossilen Rohstoffen ausgehen würde [16]. Die nähere Betrachtung zeigt jedoch, daß dabei für die menschliche Arbeitskraft in der Landwirtschaft nur der physiologische Energiebedarf (für die Ernährung) in Ansatz gebracht wurde. Die Fehlerhaftigkeit dieses Ansatzes ist offensichtlich, da der produktiv arbeitende Mensch außerhalb des Produktionsprozesses mehr Energie verbraucht als in seiner Nahrung enthalten ist. Nimmt man den Quotienten von nichtproduktiv verbrauchter Energie zu produktiv Beschäftigtem[3] als groben Schätzwert für den allgemeinen Energieverbrauch der menschlichen Arbeitskraft an, dann ergibt sich eine signifikante Veränderung der gesamten Bilanz (Abb. 2). Es ergibt sich ein neues Bild für den Vergleich mit der mikrobiellen Eiweißproduktion. Gleichzeitig ist aus dieser energetischen Bilanz die Richtung erkennbar, in der eine Intensivierung der mikrobiellen Eiweißerzeugung vorangetrieben werden muß, soll sie ökonomische Vorteile gegenüber der alternativen landwirtschaftlichen Produktion erbringen. Die Aufgabe der technologischen Forschung dazu ist, diejenigen Verfahrensstufen innerhalb des betrachteten Produktionsverfahrens zu erkennen, bei denen die angewandten technischen Maßnahmen das naturwissenschaftlich definierte Potential der ihnen zugrunde liegenden Vorgänge (Reaktion, Stoffübergang o. ä.) am wenigsten ausschöpfen, oder mit anderen Worten: deren durch die Qualität der angewandten technischen Maßnahmen bestimmte Ist-Zustände am weitesten von den durch naturwissenschaftliche Größen gegebenen Maximalzuständen entfernt sind. Die damit verbundene Aufgabe der technologischen Forschung ist die Bestimmung dieser Maximalzustände. Aus der Differenz zwischen Maximalzustand und Istzustand der Vorgänge in einer Prozeßstufe ergibt sich der Spielraum, der der technologischen Forschung bleibt, um durch Verbesserung oder Neueinführung technischer Maßnahmen die natürlich gegebenen Reserven in der betrachteten Prozeßstufe auszuschöpfen, wobei die Aufrechterhaltung der Proportionalität zwischen technischer und ökonomischer Verbesserung

[2] Die Aufwendungen für die Kohlenstoffquelle enthalten den minimalen Verbrauchskoeffizienten für das entsprechende Substrat/Organismen-Gemisch. Eventueller Mehrverbrauch geht in die Aufwendungen für den Betrieb ein, um die Größe als unbeeinflußbar durch den Produzenten gelten zu lassen.

[3] Diese Zahlenangaben stammen aus dem gleichen Zeitraum, für den die ursprüngliche Berechnung des Energieverbrauchs der Maisproduktion erfolgte [17, 18].

beachtet und kontrolliert werden muß. Das Vorgehen auf diese Weise nennen wir Maximumkonzept [19]. Wir betrachten es als ein generelles Prinzip in der technologischen Forschung. Nicht immer fallen die am weitesten vom Maximalzustand entfernten Verfahrensstufen mit den naturwissenschaftlich und technologisch am stärksten beachteten und bearbeiteten zusammen. Ziel der Strategie der technologischen Forschung muß deshalb sein, das ökonomisch Notwendige zum naturwissenschaftlich und technologisch Interessanten zu erheben.

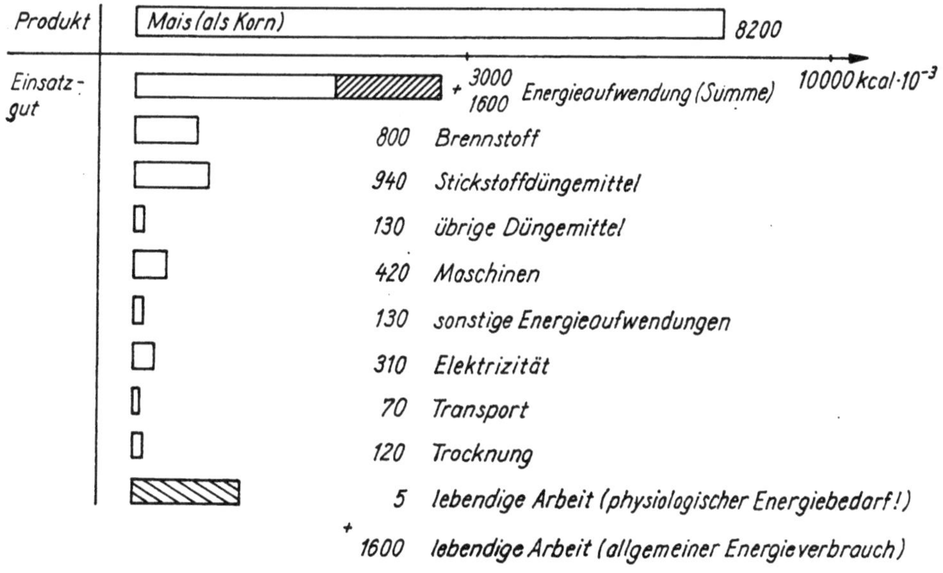

Abb. 2. Energie-Bilanz der Maisproduktion unter industriellen Bedingungen (1970), nach PIMENTEL et. al., Science 1973

Schwerpunkte der technologischen Forschung zur Ökonomisierung der mikrobiellen Eiweißsynthese aus Kohlenwasserstoffen

Für die mikrobielle Eiweißsynthese im allgemeinen erweisen sich als ökonomisch von besonderer Bedeutung [20]:

— die Gewinnung von mehr Produkten als nur Eiweiß
— die Vereinfachung der Aufarbeitungsprozesse zum Finalprodukt
— die Verbesserung der Produktivität und Ausbeute bei der Fermentation

Erzeugung weiterer Produkte

Die Gewinnung von mehr Produkten als nur Eiweiß resultiert aus der heterogenen chemischen Zusammensetzung des gebildeten Produkts und aus der Unvollständigkeit der biochemischen Umsetzung des Energieinhalts des Rohstoffs in das eigentliche Zielprodukt. Sie wird erschwert durch die Strukturierung des gebildeten Primärprodukts der mikrobiellen Biomasse. Besonders sinnfällig wird die Notwendigkeit der Gewinnung weiterer Produkte bei der mikrobiellen Eiweißsynthese aus Dieselkraftstoff (Abb. 3). Hierbei wird der heterogene Rohstoff nur unvollständig genutzt (nur der n-Alkan-Anteil) und das Verhältnis von durchgesetztem Rohstoff zu Zielprodukt beträgt etwa

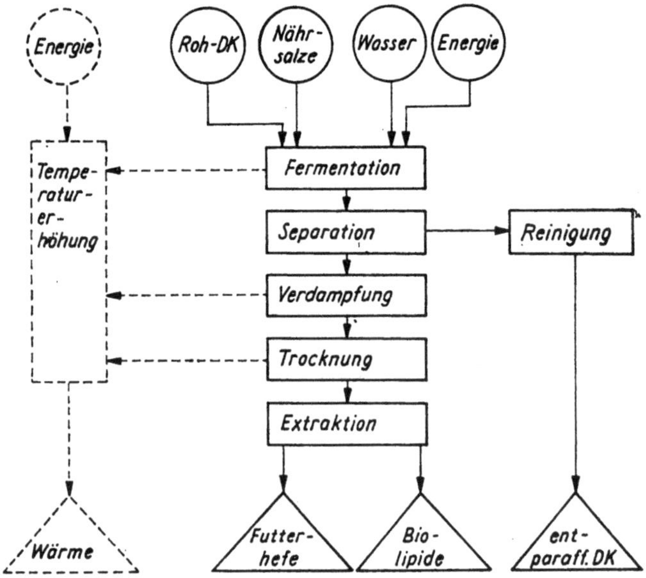

Abb. 3. Prinzipschema der Dieselkraftstoffverhefung

10:1. Der entparaffinierte Dieselkraftstoff stellt als Komponente von Winter-DK ein Produkt dar, dessen Gebrauchswerterhöhung auch bei geringer preislicher Berücksichtigung infolge der Menge einen erheblichen ökonomischen Effekt auf das Gesamtverfahren bewirkt. Auch höherwertige Verwendungszwecke sind bereits bekanntgeworden [21].
Die Konvertierung von Dieselkraftstoff zu mikrobiellem Futtereiweiß erfordert eine vollständige Trennung des Dieselkraftstoffs von den gebildeten Hefezellen. Diese Trennung wird zunächst grob mechanisch bewirkt, endgültig durch eine Lösungsmittelextraktion. Die vollständige Trennung kann nur durch Lösung der Protein-Lipid-Verbunde in der Zelle erfolgen [22, 23]. Diese Trennung gewährleistet den Abtransport der restlichen Kohlenwasserstoffe aus der Zelle, befördert aber auch alle Lipide und lipo-

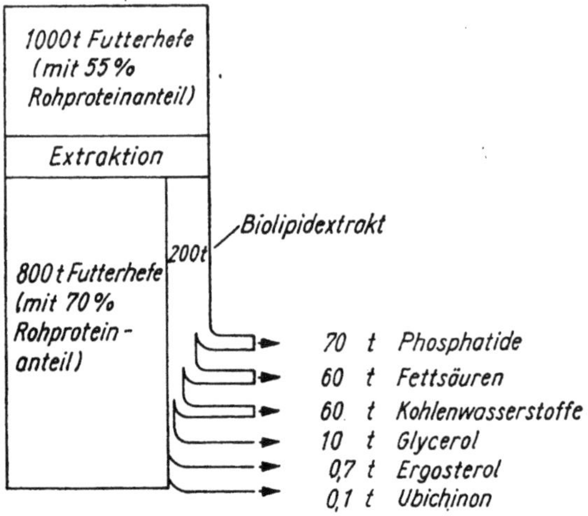

Abb. 4. Nebenprodukte der mikrobiellen Futterhefeproduktion aus Erdöldestillat

philen Bestandteile der Zellsubstanz in den von den Zellen mechanisch abtrennbaren Extrakt. Auf diese Weise entsteht nach Abdestillation des Lösungsmittels ein Produkt, das sich durch eine starke oberflächenaktive Wirkung auszeichnet und auch die Frage nach einer weiteren Auftrennung zur Gewinnung von Wertstoffen aufzuwerfen gestattet (Abb. 4) [12]. Es liegen eine Reihe von gesicherten Anwendungsmöglichkeiten dieses Extrakts vor, die den Extrakt nicht nur schlechthin unterzubringen, sondern sinnvoll und mit ökonomischem Vorteil für das Gesamtverfahren einzusetzen gestatten (Tab. 4) [24]. Die Auftrennung des Extrakts in Komponenten wurde ebenfalls in Angriff genommen, um für anspruchsvollere Einsatzgebiete auch homogene Fraktionen und Produkte anbieten zu können. Ihre Einsatzmöglichkeit kann ebenfalls als nachgewiesen gelten (Tab. 5) [25].

Tabelle 4. Verwertungsmöglichkeiten für Biolipide

Gebiet	obere Grenze in %	Wirkung
Fluoritflotation	0,04	Sammler
Lackherstellung	2	Verhinderung Pigmentabsetzung
Ölheizung	0,05	Verbrennungsförderer
Straßenbau	1	Haftungsverbesserung zwischen Bitumen und Splitt
Düngemittelherstellung	0,3	Antibackmittel
PSM-Herstellung	5	Emulgierung, Dispergierung
Bauindustrie	30	Formentrennung
Geologische Bohrungen	5	Oberflächenaktive Wirkung
Erdölverarbeitung	10	Bitumen mit verbesserten Eigenschaften
Landwirtschaft	10	in Verbindung mit Bitumen zur Bodenverbesserung
Umweltschutz	10	Emulgierung von ausgelaufenem Erdöl
Fermentation	0,1	Sauerstoffeintrag, Hefe/Öl-Trennung

Tabelle 5. Verwertungsmöglichkeiten für Fraktionen

Rohphosphatide	Straßenbau Pflanzenschutzmittelherstellung Lackindustrie
Reinphosphatide	Substitution pflanzlicher Lecithine in verschiedenen Industriezweigen
Fettsäuren	Fluoritgewinnung Lackindustrie Abwasserreinigung Fungizidherstellung
Ergosterol	Rohstoff für Vitamin D_2 Rohstoff für Steroidhormone
Ubichinone	Pharmakologie Wachstumsregulator für Mycobakterien

Als technisch nicht in hinreichendem Maß gelöst stellt sich heute die Nutzung der aus der biochemischen Umsetzung herrührenden Wärme dar, die als das weitaus größte potentielle Nebenprodukt der mikrobiellen Eiweißsynthese betrachtet werden muß (Abb. 5) [12]. Ihre Abführung über Kühlkreisläufe erfordert heute Aufwendungen, die über 10% der Gesamtaufwendungen liegen. Der minimale Wärmeanfall ist aus theoretischen Berechnungen und experimentellen Ergebnissen bekannt [26, 27], ebenso das Temperaturniveau der freiwerdenden Wärme [28] und die möglichen energetischen Wirkungsgrade bei der Transformation der Wärme auf höhere Temperaturen [29]. Sie erlauben heute nur eine Nutzung der biochemisch erzeugten Wärme unter Einsatz einer bedeutend größeren extern zugeführten Energiemenge.

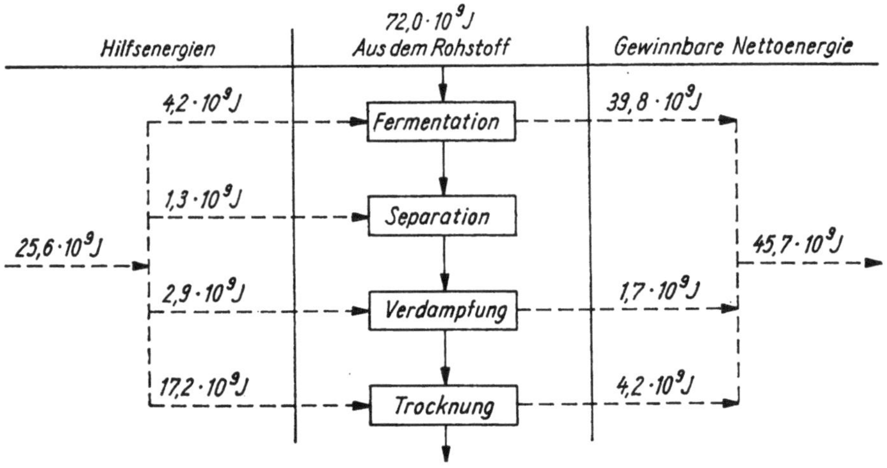

Abb. 5. Energieflußbild bei der Gewinnung von 1 t Futterhefe auf Erdöldestillat

Vereinfachung der Aufarbeitungsprozesse

Der Vereinfachung der Aufarbeitungsprozesse in technisch-mikrobiologischen Produktionen wird seit einigen Jahren erhöhte Aufmerksamkeit gewidmet, um sowohl investitionsseitig als auch betriebsseitig die Aufwendungen zu senken. Für die mikrobielle Eiweißsynthese hat die übliche stufenweise Wasserabtrennung (mechanisch-thermisch-thermisch) [30] einen erheblichen Einfluß auf die Betriebs-, Investitions- und Erhaltungsaufwendungen. Letztendlich ist hier seit den 30er Jahren kein wesentlicher Fortschritt erreicht worden, weil keine neuen naturwissenschaftlichen Erkenntnisse über die kolloide Natur der Zellbegrenzungen und ihrer Beeinflußbarkeit hinzugekommen sind. Bei der Eiweißsynthese auf Dieselkraftstoff kommt zu der Wasserabtrennung noch die Ölabtrennung hinzu. Die Anwesenheit von Öl gibt jedoch andererseits die Möglichkeit, die partiell hydrophoben Hefezellen in einer Ölphase anzureichern und von der Hauptmenge der wäßrigen Phase einfach zu dekantieren. Letztendlich wird dabei auf der Basis einer in der Zelle vorhandenen chemischen Energie mechanische Energie eingespart. Die partielle Hydrophobizität der Hefezelle kann weiter ausgenutzt werden. Läßt man die dekantierte Öl/Hefe/Wasser-Suspension an apolaren Polymeroberflächen entlanglaufen, so erfolgt eine Umorientierung der Bindungskräfte des Öls von den Hefen zu den hydrophoberen Polymeren, eine Koaleszenz der Öltropfen setzt ein und erlaubt ihre Abtrennung. Dieser Prozeß kann durch die Anwesenheit oberflächenaktiver Verbindungen beschleunigt werden. Dafür eignen sich auch die in der Extraktion ab-

getrennten Biolipide [31]. Damit ergibt sich eine Einsparung von mechanischer Energie sowie von Investitionsaufwendungen durch den Ersatz hochkomplizierter Separatoren durch einfache Röhrensysteme (Abb. 6).

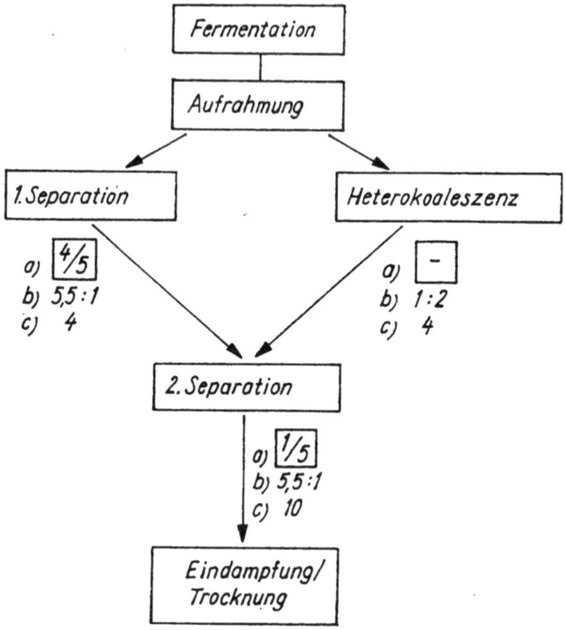

Abb. 6. Vergleich Heißseparation-Heterokoaleszenz in der Primärtrennung des 4-Phasen-Systems der ED-Konvertierung

a) Separatoreneinsatz
b) HTS: Öl } Qualität der Stoffströme
c) HTS % }

Steigerung der Produktivität in der Fermentation

Für die Behandlung der Problematik der Produktivitätssteigerung konnte von uns eine mathematische Formulierung des Maximalzustandes der Fermentation aufgestellt werden:

$$\dot{x}_{max} = \mu_{max} x_{max}$$

$\dot{x}$ = Produktivität in $kgm^{-3}\,h^{-1}$
μ = spezifische Wachstumsgeschwindigkeit in h^{-1}
x = Zellkonzentration in kgm^{-3}

Ihr lagen Studien [12] zugrunde, die zum Nachweis der Gültigkeit von Michaelis-Konstanten in Wachstumsprozessen bei hohen Zellmassekonzentrationen durchgeführt wurden. Es zeigte sich, daß bei geeigneter Wahl der Fermentationsbedingungen sehr kleine Konstanten resultieren, so daß die aus mathematischer Sicht nicht unproblematische Gleichung für die Berechnung der Maximalproduktivitäten genutzt werden kann. Für μ_{max} gelten dann die für die spezifische Organismen/Substrat-Kombination erreichten günstigeren Werte, für x_{max} die aus Viskositätsmessungen ermittelten Grenzkonzentrationen an Biomasse, bei denen noch kein signifikanter Einfluß auf die Stoff-

übertragungseigenschaften des Mediums erkannt werden kann. Beide Grenzwerte besitzen nicht die Qualität von Stoffkonstanten, es sind abgeschätzte Bereiche [32]. Sie erlauben jedoch die Angabe eines Bereichs der maximalen Produktivität, die für die Erdöldestillatverhefung[4] bei etwa 10 $kgm^{-3}\,h^{-1}$ für die Methan- und n-Paraffinverhefung noch über 25 $kgm^{-3}\,h^{-1}$ liegen [12].

Obwohl gerade bei der Erdöldestillatverhefung die Stoffübergänge der n-Alkane aus der öligen in die wäßrige und aus der wäßrigen in die feste (Mikroorganismen)-Phase augenfällig werden, ist doch in weiten Bereichen auch hier der Sauerstoffübergang aus der Gas- in die wäßrige Phase der produktivitätsbegrenzende Schritt in realen Systemen. Die Verbindung zwischen maximaler Produktivität und adäquatem Sauerstoffübergang herzustellen, ist deshalb das Ziel der technischen Maßnahme. Der Zusammenhang ist über den spezifischen Sauerstoffverbrauchskoeffizienten gegeben, der minimal sein muß, wenn die Sauerstoffübergangsgeschwindigkeit, für die externe Energie aufzuwenden ist, den kleinstmöglichen Wert annehmen soll:

$$\dot{x}_{max} = \frac{1}{\alpha_{x\,min}^{O_2}} \cdot \dot{c}_{adäqu.}$$

$\alpha_x^{O_2}$ = spezifischer Sauerstoffverbrauchskoeffizient in $kg\,kg^{-1}$
$\dot{c}$ = Sauerstoffübergangsgeschwindigkeit in $kgm^{-3}\,h^{-1}$

Wegen der Proportionalität aller Verbrauchs- und Bildungskoeffizienten, die Ausdruck der Stöchiometrie der biochemischen Umsatzgleichung ist, hat das Streben nach dem

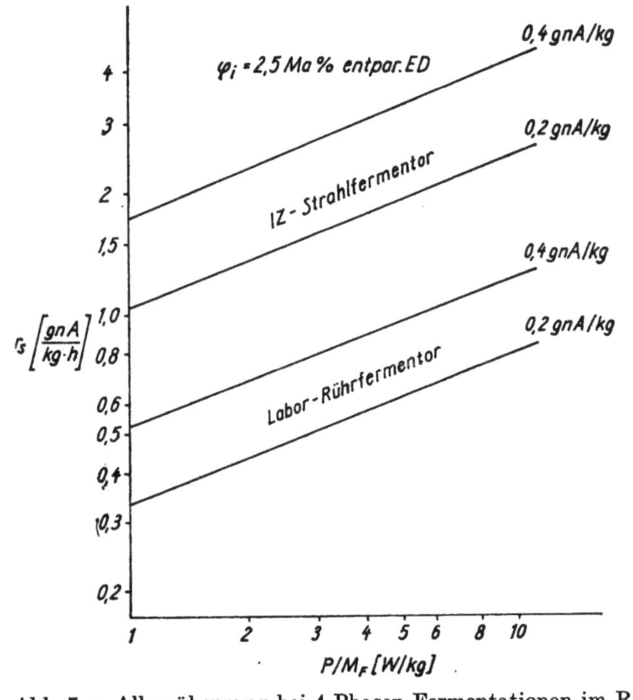

Abb. 7. n-Alkanübergang bei 4-Phasen-Fermentationen im Rührfermentor und IZ-Strahlfermentor

[4] Bei der Erdöldestillatverhefung muß berücksichtigt werden, daß es sich nicht um ein reines, sondern um ein verdünntes Substrat handelt.

minimalen Sauerstoffverbrauchskoeffizienten gleichzeitig auch Auswirkungen in Richtung minimalen Rohstoffverbrauchs und minimaler Wärmeerzeugung [12].
Für die Konstruktion geeigneter Sauerstoffübertragungseinrichtungen (Fermentoren) ist mit dem Rohrreaktorsystem die ideale Lösung vorgegeben worden, die kinetisches Verhalten der Hauptreaktionen (autokatalytische Art 1. Ordnung für die Zellvermehrung im Stundenbereich; 1. Ordnung für den Sauerstoffübergang im Sekundenbereich) und Strömungsverhalten des Reaktors (ideales Mischgefäß im Stundenbereich, ideales Strömungsrohr im Sekundenbereich) zeitmäßig in Übereinstimmung gebracht hat [33]. Damit wurde der Weg frei zu einer Erhöhung der Produktivität um eine Größenordnung, was sich in den Investitions- und Erhaltungskosten erheblich niederschlägt. Das auf empirischer Basis zur technischen Reife gebrachte Tauchstrahlsystem der IZ Böhlen nähert sich mehr als die meisten der anderen bekannten Systeme dieser theoretischen Lösung. Es konnte nachgewiesen werden, daß auch der n-Alkan-Übergang im Tauchstrahlsystem günstiger verläuft als im klassischen Rührreaktor (Abb. 7) [34].

Steigerung der Ausbeute in der Fermentation

Die minimalen Verbrauchskoeffizienten lassen sich auf Grund thermodynamischer und biochemischer Überlegungen genauer bestimmen als die maximalen Produktivitäten [12]. Aus dem Abstand zu den erreichten Werten kann der Spielraum leicht ermittelt werden, der für Intensivierungen noch zur Verfügung steht (Tab. 6) [35].

Tabelle 6. Verbrauchskoeffizienten für die Biomassebildung aus Kohlenwasserstoffen [g/g]

CH_4	theoretisch	0,90
	erreicht	1,16
$C_{16}H_{34}$	theoretisch	0,75
	erreicht	0,90
DK	erreicht	1,10

Einer der wesentlichsten Fortschritte in den letzten Jahren war die Einführung der dynamischen Prozeßführung in die kontinuierlich betriebene mikrobielle Eiweißsynthese [36, 37]. Es konnte an diskontinuierlichen und kontinuierlichen Wachstumskurven synchronisierter Hefekulturen festgestellt werden, daß der Bedarf an Substrat in der Einzellphase und Sproßphase verschieden hoch war und daß der auftretende Überschuß infolge gleichmäßigen Substratangebots in beiden Phasen veratmet wurde. Das heißt,

Tabelle 7. Dynamische Prozeßführung mit niedrigerem mittleren ED-Angebot im Vergleich zur stationären Fermentation

Prozeßführung	ED-Anteil [Ma.-%]	$\dot{x}$ [g/kg · h]	$\alpha_x^{O_2}$ [g/g]	α_x^s [g/g]	α_x^w [kJ/g]
stationär	20	2,64	2,4	1,2	8,5
dynamisch	13,5	2,6	2,0	1,05	6,8

Prozeßführung	Stockpunkt des ED [°C]	Lipide [%]	Kohlenhydrate [%]	Rohprotein [%]	Asche [%]
stationär	−28	22	15	57	6
dynamisch	−40	17	12	64	6

bei nichtsynchronisierten Kultivierungen, in denen jeweils etwa gleiche Anteile von Zellen in den unterschiedlichen Zuständen vorliegen, muß Substrat unnötig veratmet, Sauerstoff unnötig mehr verbraucht und Wärme unnötig mehr erzeugt werden. Durch Synchronisierung der Kulturen und zeitgerechte Dosierung des Substrats, was einfach zu bewerkstelligen ist, wurde eine erhebliche Verbesserung nicht nur der Verbrauchskennziffern, sondern auch der Qualitäten der Biomasse und des entparaffinierten Dieselkraftstoffs erreicht (Tab. 7).

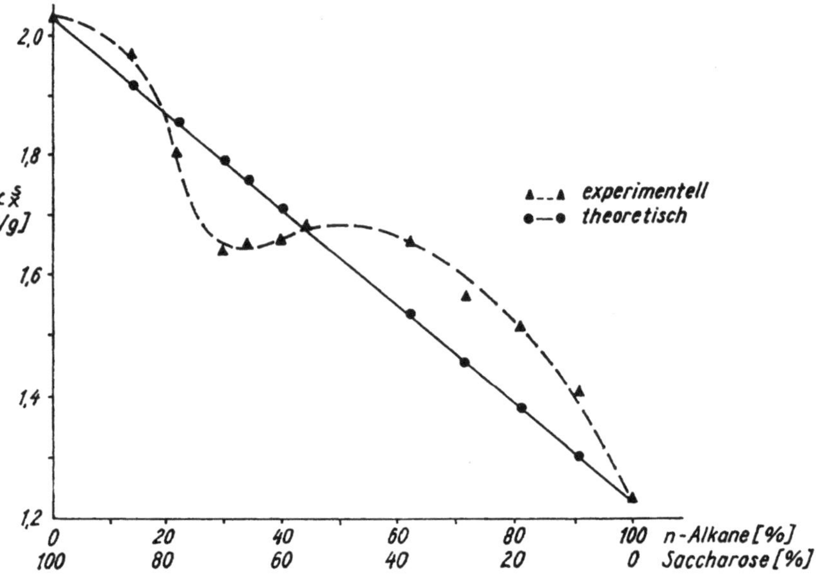

Abb. 8. Abhängigkeit der Substratverbrauchskoeffizienten vom Mischungsverhältnis der C-Quellen

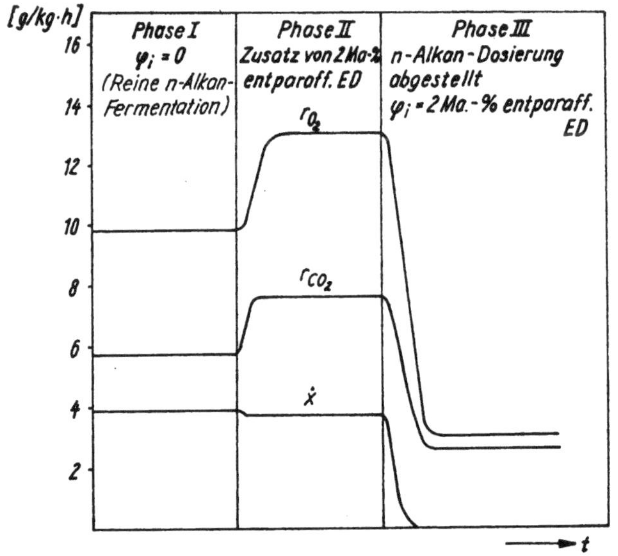

Abb. 9. Veränderung des spezifischen O_2-Verbrauchs in Gegenwart einer Nicht-n-Alkan(KW)-Phase

Von grundsätzlicher Bedeutung erwiesen sich auch Überlegungen zum optimalen Kohlenstoff-Energie-Verhältnis für die Zellmassesynthese in Kohlenwasserstoffen und anderen Substraten [38]. Sie führten zu Experimenten mit Mischungen von Saccharose und n-Alkanen, die für bestimmte Verhältnisse beider Substrate eine günstigere Ausnutzung beider Kohlenstoffquellen als Mischsubstrat bestätigen konnten als bei ihrer separaten Umsetzung (Abb. 8) [39]. Damit ist für die vollständigere Nutzung von Rohstoffen ein weites Feld praktischer Kombinationen geöffnet worden.
Bei zwangsweisen Mischungen, wie sie z. B. im Dieselkraftstoff vorliegen, werden jedoch auch negative Effekte beobachtet. Der Unterschied in den Verbrauchskoeffizienten bei Dieselkraftstoff (1,10) und n-Alkanen (0,90) konnte durch den Nachweis der Veratmung von Toluol und seinen Abkömmlingen — zumindest partiell — geklärt werden (Abb. 9) [40].

Auswahl von Mikroorganismen

Die beschriebenen Richtungen und Ergebnisse der technologischen Forschung zur ökonomischen Gestaltung mikrobieller Eiweißsynthesen müssen unbedingt um eine wesentliche ergänzt werden: die Auswahl der verwendeten Mikroorganismenkultur, die im Falle der mikrobiellen Eiweißsynthese gleichzeitig Katalysator und Produkt ist. Hier bieten sich durch die neuen Erkenntnisse der Molekularbiologie und Genetik weite Felder der Konstruktion von Organismen mit gewünschten Eigenschaften an. Es muß jedoch auch festgestellt werden, daß diese gelenkte Konstruktion methodisch noch in den allerersten Anfängen steckt und daß über ihren zweifellos großartigen Perspektiven nicht die näherliegenden und noch keinesfalls erschöpften Methoden der Selektion von Organismen aus der Natur, die Regulation ihres Stoffwechsels und die Einführung weiterer, den Eigenschaften der Organismen besser angepaßter technischer Maßnahmen vergessen werden dürfen.

Schlußfolgerungen

Das Instrumentarium zur Beurteilung und Beeinflussung der Ökonomie der mikrobiellen Eiweißsynthese aus Kohlenwasserstoffen hat sich trotz eines Rückganges der internationalen wissenschaftlichen Aufmerksamkeit für dieses Gebiet in den letzten Jahren erheblich verbreitert. Deutet man die Zeichen richtig, die vor allem in dem Fortschritt in der mikrobiologischen Industrie der UdSSR und in den Auffassungen erdölführender Entwicklungsländer zur mikrobiellen Eiweißsynthese aus fossilen Rohstoffen gesetzt wurden, dann steht eine allgemeine internationale Reanimation dieser Entwicklung bevor. Rohstoffsorgen gibt es nicht. Bei weisem Gebrauch verfügt die Menschheit noch für Jahrhunderte über Vorräte an Erdöl und Erdgas für den Einsatz in der chemischen und mikrobiologischen Industrie [41].
Es sollte mit diesem Beitrag die Gewißheit vermittelt werden, daß die mikrobielle Eiweißsynthese auf fossilen Rohstoffen durch die wissenschaftliche Erschließung der ihr innewohnenden Reserven und durch ihre industrielle Organisation wettbewerbsfähig ist und ihren Platz bei der weiteren Entwicklung der Produktivkräfte der menschlichen Gesellschaft erringen wird.

Literatur

[1] Rytschkov, R. S.: Vestnik Akad. nauk SSSR (1982) 4, 24.
[2] Arima, K.: Presentation at the „International Symposium of S.C.P." 28.—20. Jan. 1981 Paris. APRIA.
[3] Hepner, L.: GBF: 2. Symposium 1980. Mikrobielle Proteingewinnung und Biotechnologie. Stöckheim. Verlag Chemie, Weinheim. S. 1—7.

[4] Dimter, L.: Zusammenstellung der Abt. Information des Instituts für technische Chemie, Leipzig. 1982.
[5] Al Turki, A.: Presentation at the „International Symposium of S.C.P.“ 28.—30. Jan. 1981. Paris. APRIA.
[6] Leysen, R.: Chem. and Ind. (1982) 13, 428.
[7] Grigorjan, A. N.: Mendelejew-Journal **17** (1972) 512.
[8] Takahashi, H.: Vortrag anläßl. 30. Generalversammlung der IUPAC. 2.—10. Sept. 1979. Davos.
[9] Hamer, G.: Presentation at the „VI. Internat. Fermentation-Symposium“ 20.—25. 7. 1980. London, Canada.
[10] Bauch, J., Bohlmann, D., Koslowa, L. I., Meschtschankin, G. I., Roschkowa, M. I., Triems, K., Ringpfeil, M., Sobek, K.: Vortrag. XII. Internat. Kongreß für Mikrobiologie. Sept. 1978, München.
[11] Tschepigo, S. V., Gradowa, N. B., Wrobjowa, W. S., Piterskij, W. S.: Abhandlungen der AdW der DDR. 1975. Nr. 1N, S. 45.
[12] Ringpfeil, M.: Proceed. 10th World Petroleum Congress, Bukarest 1979. Vol. 4, pp 377.
[13] Frohn, W.: Neues Deutschland v. 24. 6. 1982.
[14] Keil, G., Apelt, E., Friedrich, G., Grübler, J.: Mitteilungsblatt der Chem. Gesellschaft der DDR. (1982) 6, 127.
[15] Pimentel, D., Huro, L. E., Bellotti, A. C., Forster, M. B., Oka, L. N., Sholes, O. D., Lohitman, R. I.: Science 182 (1973) 443.
[16] Böhme, H., Keil, G., Philipp, B., Ringpfeil, M.: Sitzungsberichte der AdW der DDR. 1978. Nr. 19N. S. 3—34.
[17] Balzhiser, R. E.: Chem. Engng. **89** (1982) 74.
[18] Meyers Neues Lexikon, Leipzig. 1976. S. 417ff.
[19] Ringpfeil, M.: Abhandlungen der AdW der DDR. 1982. Nr. 2N. S. 29—42.
[20] Ringpfeil, M.: Trainingskurs Biotechnologie der UNESCO. Leipzig. 1982.
[21] Hieke, W., Kockert, M., Gentszch, H., Bauch, J., Bohlmann, D.: Chem. Technik **34** (1982) 418.
[22] Ringpfeil, M.: Sitzungsberichte der AdW der DDR. 1980. Nr. 16N. S. 5—12.
[23] Beck, D., Biedermann, W., Heinze, G., Martin, S., Schmidt, J., Malzahn, J., Ringpfeil, M.: Sitzungsberichte der AdW der DDR. 1980. Nr. 16N. S. 144—161.
[24] Voigt, B., Seidel, H., Müller, H., Beck, D., Ringpfeil, M., Riedel, M., Bauch, J., Gentzsch, H., Bohlmann, D.: Chem. Technik **31** (1979) 409.
[25] Voigt, B.: Acta Biotechnol. **2** (1982) 155.
[26] Glombitza, F., Heinritz, B., Rogge, G., Bley, T.: Ztschr. allg. Mikrobiologie **22** (1982) 535.
[27] Heinritz, B., Stoll, P., Glombitza, F.: Acta Biotechnol. **3** (1983) 85.
[28] Ringpfeil, M., Triems, K.: Vortrag anläßl. 4. Fachtagung „Verarbeitung von Erdöl und Erdgas“. Frankfurt/O. 25. 4. 1980.
[29] Stephan, W., Wiesner, A.: Unveröff. Ergebnisse.
[30] Ringpfeil, M., Vetterlein, G., Schneider, J., Franke, G.: Presentation at the „International Symposium of S.C.P.“ 28.—30. Jan. 1981. Paris. APRIA.
[31] Pickert, H., Konieczny, B.: Unveröffentl. Ergebnisse.
[32] Böhme, H., Keil, G., Philipp, B., Ringpfeil, M.: spectrum (1978) 7, 18 und spectrum (1978) 8, 28.
[33] Ringpfeil, M.: Presentation at the „VI. Internat. Fermentation-Symposium“, London, Canada. 20.—25. 7. 1980.
[34] Otto, G., Grosse, M., Thamm, R.: Unveröffentl. Ergebnisse.
[35] Babel, W.: Abhandlungen der AdW der DDR. 1982. Nr. 2N. S. 183.
[36] Ringpfeil, M.: Vortrag anläßl. der Vorstellung von Akademiemitgliedern zum Leibniztag 3. 7. 1981.
[37] Heinritz, B., Stichel, E., Bley, T., Rogge, G., Glombitza, F.: Ztschr. allg. Mikrobiol. **21** (1981) 581.
[38] Babel, W.: Ztschr. allgem. Mikrobiologie **19** (1979) 671.
[39] Richter, K.: Poster. Leipziger Biotechnologie-Symposium 1982.
[40] Otto, G., Serfas, O.: Unveröffentl. Ergebnisse.
[41] Alexandrow, A.: Neues Deutschland v. 21./22. 4. 1979

Acta Biotechnol. 3 (1983) 3, 241–249

Характеристика углеводородокисляющих дрожжей Особенности их роста и биосинтетических процессов

Н. Б. Градова, Э. М. Диканская, З. Н. Робышева, Г. С. Родионова, М. Б. Биттеева, А. И. Заикина

Всесоюзный научно-исследовательский институт биосинтеза белковых веществ, 109004 г. Москва. Б. Коммунистическая 27

Symposium on Biotechnology 1982, Leipzig, G.D.R.

Summary

In this review are described results of the research in the Soviet Union in the field of selection and physiology of yeast strains for the production of SCP on the base of petroleum hydrocarbons.
The most efficient yeast strains were selected by cultivation on enrichment medias. Most of them belong to a limited number of the species of the genus *Candida*.
The stability of the industrial interesting properties of yeasts also was demonstrated during a continuous long-time cultivation.

Zusammenfassung

Es wird ein Überblick über Forschungsergebnisse in der Sowjetunion auf dem Gebiet der Selektion und Physiologie von Hefestämmen zur Gewinnung von Einzellerprotein aus Erdölkohlenwasserstoffen gegeben.
Die produktivsten Hefestämme wurden durch Kultivierung auf Anreicherungsmedien selektiert. Die meisten gehörten zu wenigen Arten der Gattung *Candida*.
Besonders hervorgehoben wird die Beständigkeit der industriell interessanten Eigenschaften von Hefen auch während einer kontinuierlichen Langzeitkultivierung.

При создании производства белка одноклеточных одной из основных задач является создание технологии, способной обеспечить условия для достижения максимальной продуктивности роста микроорганизма и для разработки процессов биосинтеза практически ценных компонентов клетки, обусловленных их генотипом. Решение этой задачи предусматривает правильный выбор микроорганизма-продуцента и глубокое знание его физиологии.

В настоящем сообщении представлены обобщенные результаты исследований в области селекции и физиологии штаммов дрожжей, используемых для получения белка одноклеточных из углеводородов нефти.

Селекция углеводородокисляющих дрожжей. Работы по интродукции из природы дрожжей, способных усваивать углеводороды в качестве единственного источника углерода и энергии, позволили отобрать высокопродуктивные штаммы для получения белка одноклеточных.

Использование для выделения из природных образцов — метода прямого посева на сложную питательную среду (сусло — агар) с последующим выделением чистых культур показало, что способность усваивать углеводороды в качестве единственного источника углерода и энергии широко распространена среди большого числа видов дрожжей рода *Candida*, *Rhodotorula* и некоторых видов *Trichosporon* и *Torulopsis*.
Наибольшее число выделенных штаммов относилось к роду *Candida*. Среди дрожжей этого рода обнаружены широкие различия по эффективности усвоения углеводородов и отобраны наиболее продуктивные штаммы.
Соответствующие данные имеются и в литературе [1].

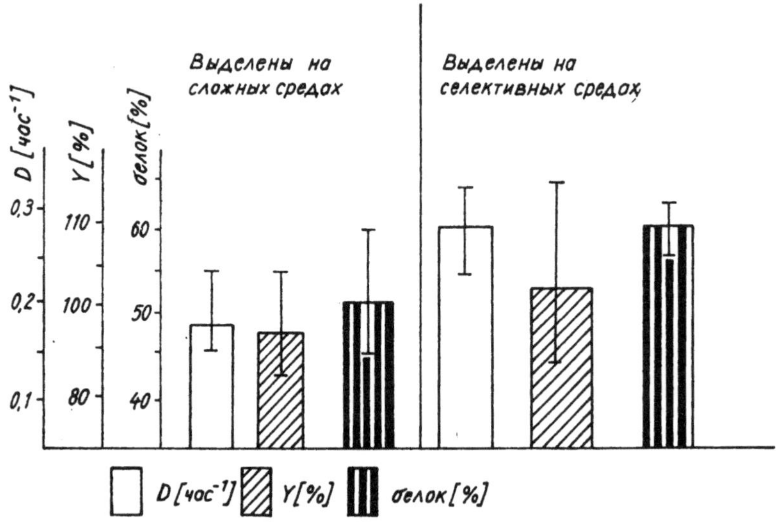

Рис. 1. Показатели роста углеводородокисляющих дрожжей, селекционированных разными методами

Применение для выделения и селекции углеводородокисляющих дрожжей методов, основанных на естественном отборе, а именно метода накопительных культур в сочетании с методом автоселекции в процессе непрерывного культивирования на селективных средах, позволило отобрать из многомерной экологической ниши углеводородусваивающих дрожжей, составляющие ее параметрические ниши — культуры, характеризующиеся высокой скоростью роста, высоким экономическим коэффициентом усвоения углеводородов, а также растущие в разных интервалах температур (мезофильные и термотолерантные) [2].
Идентификация штаммов показала, что все они относятся к ограниченному числу видов рода *Candida*: *C. parapsilosis*, *C. maltosa*, *C. sake*, *C. tropicalis* или близки к ним.
Для одного из этих штаммов (*C. parapsilosis*) выявлена совершенная форма и он отнесен к *Lodderomyces elongisporus* [3].
Люминисцентно-серологическим методом с использованием сывороток, полученных к целым клеткам, показана общность антигенного состава селекционированных активных штаммов.
При использовании автоселекции, в результате которой отбираются формы, отличающиеся интегральной приспособленностью к условиям культивирования, получены штаммы дрожжей, характеризующиеся суммой промышленно-ценных свойств: высокой скоростью роста, продуктивностью и высоким содержанием белка в биомассе (рис. 1.).

Отношение к индивидуальным алканам. Углеводородокисляющие дрожжи способны использовать в качестве единственного источника углерода и энергии алканы с длиной углеродной цепи от C_9-C_{10} до $C_{28}-C_{30}$. С наибольшим экономическим коэффициентом усваиваются алканы $C_{16}-C_{18}$. При уменьшении длины углеродной цепи алканов снижается экономический коэффициент усвоения субстрата. Большая эффективность усвоения алканов с длиной цепи $C_{16}-C_{18}$ объясняется возможностью непосредственного включения в липидные компоненты клетки образующихся в качестве промежуточных продуктов окисления высокомолекулярных жирных кислот. Это позволяет клетке максимально использовать для конструктивного обмена энергию, высвобождающуюся при окислении алканов. В то же время жирные кислоты, образующиеся при окислении алканов с длиной углеродной цепи C_{14} и ниже подвергаются β-окислению, а высшие ненасыщенные жирные кислоты должны синтезироваться *de novo*, что приводит к затрате энергии.
Степень снижения экономического коэффициента усвоения алканов с длиной цепи ниже C_{14} и выше C_{18} зависит от штамма.

Рис. 2. Влияние скорости протока среды на продуктивность штаммов углеводородокисляющих дрожжей

Продуктивность роста дрожжей на фракциях углеводородов, н-парафинах и нефтяных дистиллятах, определяется соотношением алканов с определенной длиной углеродной цепи.

Оптимальные условия роста. Селекционированные штаммы дрожжей способны развиваться в широких интервалах значения коэффициента скорости протока среды, температуры и pH. Дрожжи растут при скорости протока среды Д от 0,1 до 0,7 $час^{-1}$. Однако максимальная продуктивность и минимальный удельный расход углеродного субстрата на единицу биомассы достигается в узком интервале значений скорости протока среды, различном для разных штаммов (рис. 2).
Мезофильные штаммы дрожжей растут без снижения продуктивности при температуре от 32 до 35°. При температуре 36—37° продуктивность снижается, но в различной степени у разных штаммов (рис. 3).
Термотолерантные дрожжи растут без снижения продуктивности при температуре до 38—39°. Выращивание при более высокой температуре приводит к снижению экономического коэффициента усвоения углеводородов и содержания белка в биомассе (рис. 3).
Дрожжи способны развиваться при pH среды от 2,5 до 5,5. По этому признаку отмечаются большие штаммовые различия. Выделены штаммы одинаково активно усваивающие углеводороды в диапазоне pH от 3,0—3,5 до 5,0—5,5 (рис. 3).

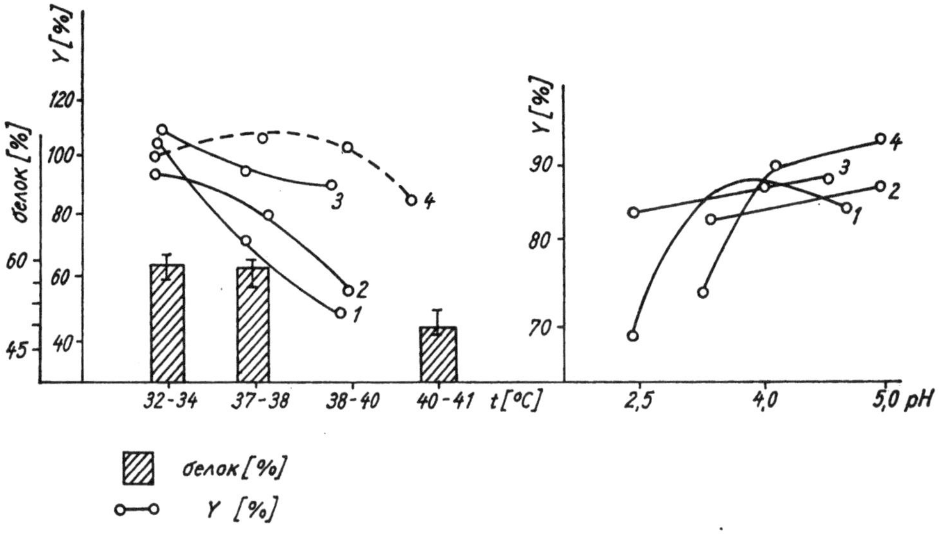

Рис. 3. Влияние температуры и pH среды на рост разных штаммов углеводородокисляющих дрожжей

Потребность углеводородокисляющих дрожжей в факторах роста. Все селекционированные активные штаммы дрожжей р. *Candida* нуждаются в биотине. Однако при развитии за счет углеводородов потребность в этом витамине ниже, чем на углеводных средах. Исследования показали [4], что степень потребности дрожжей в биотине зависит от длины углеродной цепи используемого алкана. При усвоении алканов с длиной углеродной цепи C_{16} и выше штаммы способны расти без добавок биотина, а потребность их в биотине проявляется в снижении скорости роста и выхода биомассы (рис. 4, 6).

При развитии на фракциях н-парафинов степень потребности дрожжей в биотине зависит от соотношения в них алканов с разной длиной углеродной цепи [5].

Предполагается, что уменьшение потребности дрожжей в биотине при росте на алканах выше C_{16}, очевидно, связана с тем, что часть высших жирных кислот, в синтезе которых участвует биотин, образуются в результате прямого окисления субстрата [4].

Рост углеводородокисляющих дрожжей стимулируется аминокислотами (глутаминовая кислота, аргинин). Стимуляция проявляется независимо от длины углеродной цепи алкана, используемого для роста. Стимулирующий эффект аминокислот не связан с каким-либо биотиноподобным действием, так как проявляется только в присутствии биотина в среде [5].

Продукты метаболизма углеводородокисляющих дрожжей и их влияние на рост. При развитии дрожжей на углеводородах в культуральной жидкости накапливаются продукты неполного окисления углеводородов, концентрация которых зависит от штамма и условий культивирования [6].

Дрожжи способны использовать продукты метаболизма нейтральной природы практически полностью после потребления углеводородов из среды. Органические кислоты, только частично используются дрожжами. Низкомолекулярные органические кислоты и кислоты с разветвленной цепью оказывают угнетающее влияние на рост углеводородокисляющих дрожжей даже в низких концентрациях, 0,01—0,001%. При снижении значения pH среды ингибирующее действие органических кислот усиливается [6].

Рециркуляция культуральной жидкости при культивировании дрожжей на углеводородах в зависимости от условий выращивания, с которыми связана концентрация продуктов метаболизма, может оказывать или стимулирующее, или ингибирующее действие на рост дрожжей.

По степени устойчивости дрожжей к собственным продуктам метаболизма выявлены большие различия между штаммами.

Содержание белка. Показаны большие различия в содержании белка у штаммов, выделенных из природы, от 48 до 62%. У активных углеводородокисляющих штаммов содержание белка (N × 6,25) колеблется в пределах 56—62% и не зависит от используемого источника углерода.

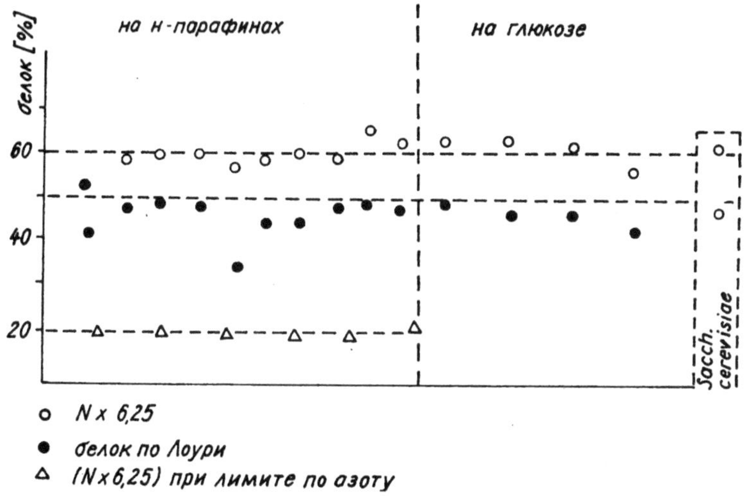

Рис. 4. Содержание белка в биомассе штаммов углеводородокисляющих дрожжей р. *Candida*

По содержанию белка в клетках и соотношению белковых и небелковых азотсодержащих соединений углеводородокисляющие штаммы дрожжей рода *Candida* практически не отличаются от сахаромицетов [рис. 4]. Вместе с тем у исследованных штаммов этот признак подвержен широким фенотипическим изменениям. Так, при лимитировании роста дрожжей азотом, или при выращивании в условиях сниженной аэрации содержание белка в клетках этих дрожжей может снижаться с 50—60% до 17—20%.

Аминограмма белка разных штаммов углеводородокисляющих дрожжей довольно однообразна. Содержание отдельных аминокислот изменяется в узких пределах. Так, содержание лизина колеблется в пределах 4,0—5,5% к сухому веществу, серусодержащих аминокислот 0,8—1,5%, а сумма незаменимых аминокислот в пределах 23,5—26,0%.

Биосинтез соединений витаминной природы. Селекционированные продуктивные штаммы углеводородокисляющих дрожжей относятся к числу активных синтетиков соединений витаминной природы. Установлено, что использование дрожжами углеводородов приводит не только к изменению потребностей в витаминах, но и к выраженным количественным изменениям процесса витаминогенеза [1, 8]. При культивировании дрожжей на углеводородных средах (рис. 5) повышается по сравнению с углеводными средами содержание в клетках флавинов, пантотеновой

и фолиевой кислот, НАД коферментов, эргостерина, но снижается содержание в них тиамина [9].

Содержание флавинов у ряда культур на средах с углеводородами возрастает почти в два раза за счет увеличения уровня ФАД [8, 13]. Из литературы известно, что при этом может меняться уровень других витаминов и коферментов в дрожжевой клетке.

Таким образом переход дрожжевой клетки к окислению углеводородов отражается на синтезе соединений витаминной природы разных групп.

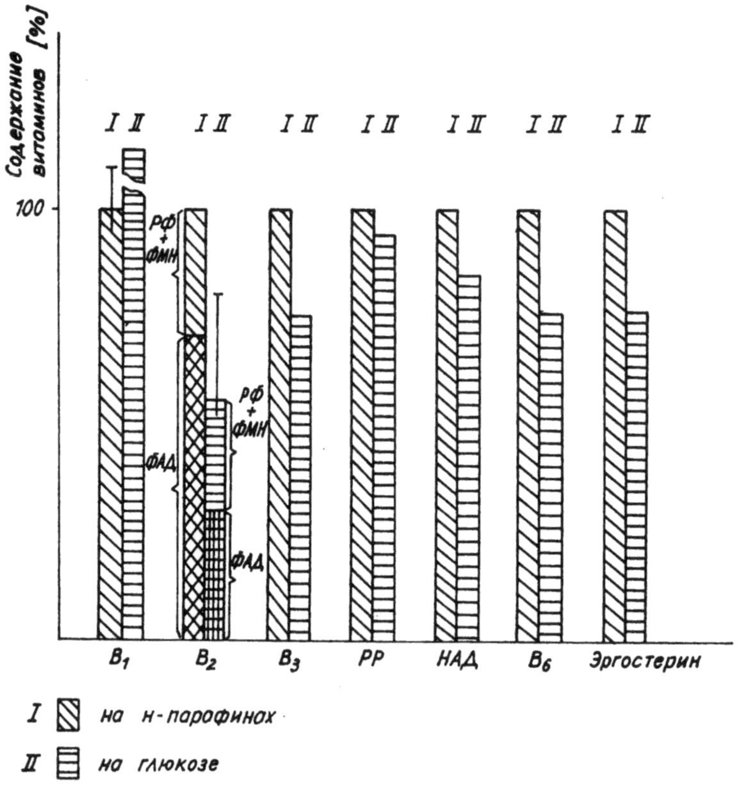

Рис. 5. Содержание витаминов в дрожжах р. *Candida* при росте на н-парафинах и глюкозе.

Изменения процессов витаминогенеза, связанные со сменой углеродного субстрата, могут быть вызваны увеличением или уменьшением роли тех или иных витаминов в метаболизме дрожжей, т. е. изменением физиологической потребности клетки в этих соединениях. Так на средах с углеводородами увеличивается обеспеченность дрожжей именно ФАД коферментом, который принимает у участие в окислении первых продуктов деградации алканов, а не ФМН [9, 13].

Однако это не единственная причина изменений процессов витаминогенеза. На примере рибофлавина показано [8], что окисление алканов наряду с повышением синтеза ФАД может вызвать у некоторых культур значительное увеличение количества витамина, выделяемого в среду, очевидно в следствии нарушения регуляторных механизмов.

Установлено, что уровень синтеза дрожжами отдельных соединений витаминной природы на углеводородных средах, как и степень потребности в биотине, зависит

от длины цепи используемых алканов [8, 9, 13]. Изменение длины углеродной цепи может вызвать комплекс изменений в росте и биосинтетических процессах углеводородокисляющих дрожжей (рис. 6).

Выявлена значительная фенотипическая изменчивость уровней витаминогенеза, не связанная с используемыми источниками углерода. Степень такой изменчивости различна для разных штаммов и разных витаминов. Значительные изменения содержания витаминов наблюдаются, в частности, при изменении концентраций азота и фосфора в среде [14].

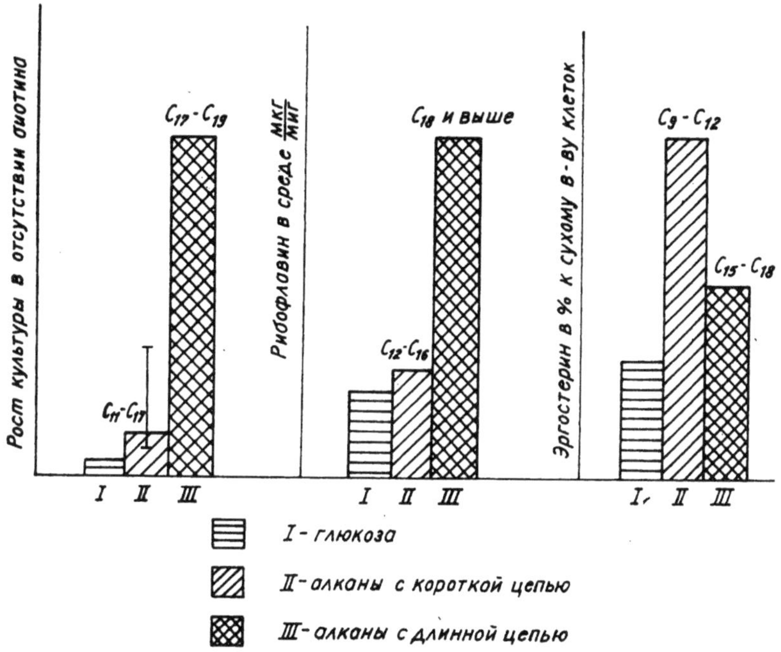

Рис. 6. Зависимость уровня витаминогенеза от углеродной цепи алканов
I — глюкоза
II — алканы с короткой цепью
III — алканы с длинной цепью

Полученные данные показывают возможности регуляции для практических целей биосинтетических процессов углеводородокисляющих дрожжей без применения генетических методов.

Химический мутагенез и генотипическая изменчивость дрожжей. Исследование действия химических мутагенов на углеводородокисляющие штаммы дрожжей р. *Candida* позволило выявить ряд особенностей этих организмов: это, — во-первых — их резистентность к химическим мутагенам, и, во-вторых, проявление стимулирующего эффекта низких концентраций химических мутагенов на их рост.

Изучение мутабильности дрожжей по морфологическим признакам, по признаку продуктивности (образование единицы биомассы в единицу времени), содержанию белка в биомассе, ацидо-, термотолерантности и снижению потребности в биотине показало, что размах генотипической изменчивости дрожжей по данным признакам различен.

Индуцированная генотипическая изменчивость дрожжей по таким признакам, как продуктивность и содержание белка в биомассе, которые контролируются поли-

генно, находится в пределах фенотипической изменчивости. Однако размах последней у полученных мутантов снижен [10, 11].

Наибольшая генотипическая изменчивость дрожжей наблюдается по признаку ацидотолерантности и наименьшая но морфологическим признакам и признаку термотолерантности.

Под воздействием химических мутагенов получена прямая мутация от гетеротрофности по биотину к автотрофности.

Исследование взаимосвязи между ингибирующим, стимулирующим и мутагенным эффектами химических мутагенов позволило установить, что по ряду признаков наибольшая мутагенная активность отмечается для стимулирующих и слабоингибирующих доз.

Непрерывная культура углеводородокисляющих дрожжей. Популяционный фенотипический анализ углеводородокисляющих дрожжей показал, что в стабильных условиях длительного непрерывного культивирования устойчиво сохраняются основные промышленно-ценные свойства штаммов, в частности, продуктивность, содержание белка в биомассе [3, 7].

Наблюдаемые изменения морфологических признаков клеток, составляющих популяцию, зависят в пределах их фенотипической изменчивости от характера изменений факторов внешней среды. Применение оптикоструктурного машинного анализа при использовании комплекса Протва-ЭВМ позволило выявить специфичность некоторых морфологических изменений. Это дает основание для разработки морфоло-физиологических методов контроля и управления процессами культивирования дрожжей.

Периодические изменения параметров непрерывного культивирования дрожжей приводят к отбору форм, характеризующихся широкой нормой реакции на изменение данных условий внешней среды.

Непрерывная культура углеводородокисляющих дрожжей в незащищенных условиях ферментации можно рассматриваться как специфическая технологенная экологическая ниша [12].

Изучение закономерностей формирования видовой и трофической структуры данной системы показало, что углеводородокисляющие дрожжи являются высшим трофическим уровнем системы. Продукты метаболизма дрожжей обуславливают формирование второго трофического уровня, поликультуру сопутствующей микрофлоры.

В целом, данная система характеризуется линейной метаболической цепью, но взаимоотношения между трофическими уровнями характеризуются не только метабиотическим типом, но и типом мутуализма, поскольку сопутствующая микрфлора может обеспечивать витаминами ауксогетеротрофные штаммы дрожжей.

Представленные результаты исследований послужили основой для разработки технологии крупнотоннажного производства получения дрожжей из н-парафинов и нефтяных дистиллятов. Эти исследования дают возможность наметить дальнейшие перспективные направления интенсификации существующих производств, а также создания новых производств по получению отдельных продуктов микробиологического синтеза на основе углеводородов.

Литература

[1] Градова, Н. Б., Диканская, Э. М., Михалева, В. В.: Использование углеводородов дрожжами, изд-во СОНТИ-микробиопром, Москва 1971 г.

[2] Gradova, N. B., Chepigo, S. V., Rodionova, Y. S., Michaleva, V. V., Bravicheva, R. N., Sustina, R. N.: Proc. 3th Intern. Spec. Symp. Yeasts, Finland, Helsinki 1973.

[3] WÜNSCHE, L., SATTLER, K., GRADÔVA, N., MEINHOLD, I., HEDLICH, R., BRENDLER, W., UHLIG, M., RODIONOVA, Y., SAIKINA, A.: Allg. Mikrobiol. **21** (1981) 469.
[4] ДИКАНСКАЯ, Э. М.: Успехи микробиологии **15** (1980), стр. 68.
[5] РОБЫШЕВА, З. Н., ОСИПОВА, В. Г., ДАВИДОВА, Е. Г., МЕЛКОНЯН, А., ГРАДОВА, Н. Б.: Метаболизм микроорганизмов и его регуляция. Тезисы докладов УI съезда Всесоюзного микробиологического общества, г. Рига, 25.—29. III. 1980 г., т. 3, стр. 85.
[6] ТРЕТЬЯКОВ, В. П., ГРАДОВА, Н. Б., РУБАН, Е. Л.: Прикл. биохим. и микроб. **6** (1970) 3, стр. 297.
[7] GRADOVA, N., BRAVICHEVA, R., KURONOVA, N., RODIONOVA, Y., SAIKINA, A.: Abstr. Vth Intern. Ferment. Symp., Berlin 1976, S. 22.
[8] ДИКАНСКАЯ, Э. М.: Биосинтез флавинов микроорганизмами. Сб. Итоги Науки, сер. ,,Биология'', М., изд. ВИНИТИ **3** (1972), стр. 5.
[9] ДИКАНСКАЯ, Э. М.: Труды УI Всесоюзного съезда микробиологического общества. Тезисы докладов, г. Рига, 1980 г.
[10] ГРАДОВА, Н. Б., ОСИПОВА, В. Г., ДАВИДОВА, Е. Г., ЧУНАЕВ, А. С., КВИТКО, К. В.: Генетика **12** (1976) 7, стр. 7.
[11] GRADOVA, N. B., ROBYSHEVA, Z. N.: Abstr. VIth Ferm. Symp. and Vth Intern. Symp. Yeasts, London, Ontario, Canada 1980, 225.
[12] ГРАДОВА, Н. Б., КОВАЛЬСКИЙ, Ю. В.: Микробиология **47** (1978), стр. 259.
[13] ДИКАНСКАЯ, Э. М., ГОРОБЦОВА, Т. А.: Микробиология **43** (1974), стр. 961.
[14] ДИКАНСКАЯ, Э. М., ДОЛГАЯ, М. Б.: В сб. ,,Лимитирование и ингибирование процессов роста и микробиологического синтеза''. Научный центр биологических исследований, г. Пущино, 1976, стр. 157.

Acta Biotechnol. 3 (1983) 3, 250

BOOK REVIEW

G. Seltmann

Die bakterielle Zellwand

VEB Gustav Fischer Verlag Jena, 1982,
1. Auflage, 207 Seiten, 60,00 M
(unter Mitarbeit von O. Kühne und R. Reissbrodt)

Die vorliegende Monografie macht es sich zur Aufgabe, die in den letzten Jahren z. T. sprunghaft gewachsenen Kenntnisse zum Aufbau, zur Struktur und zur Funktion der bakteriellen Zellwand darzustellen. Sie schließt damit in erfreulicher Weise eine seit geraumer Zeit bestehende Lücke im i. b. deutschsprachigen Schrifttum. Der Stoff wird in 8 Kapiteln mit klarem Aufbau bei guter Übersichtlichkeit dargelegt. Tabellarische Übersichten, einprägsame schematische Darstellungen und übersichtliche Grafiken geben dem Leser damit Mittel zur Hand, an denen er sich rasch informieren, seine Kenntnisse auffrischen bzw. vervollständigen sowie wichtige Hinweise über Isolation, Reinigung und Charakterisierung von Einzelkomplexen bzw. Einzelkomponenten finden kann. Fragen der Biosynthese der Substanzklassen, deren Regulation und Beeinflussung stellen einen wichtigen Bestandteil der Monografie dar und vermitteln dem Leser, dessen Interesse mehr auf therapeutischem als auf biochemischem Gebiet liegt, wichtige Ansatzpunkte bei z. B. der Bekämpfung pathogener Keime. Modelle für Membran und Zellwand werden ebenso behandelt wie Fragen der funktionellen Bedeutung der Zellwandkomponenten und der Zellwand, darunter so wichtige Fragen wie die Bedeutung der Zellwand für die Pathogenität und Virulenz, das Haften von Bakterien auf Oberflächen, der Wechselwirkung der Zellwand mit Bakteriophagen u. a. m. Jedes der 8 Kapitel wird mit weiterführender Literatur untersetzt. Ergänzungen am Schluß beinhalten Publikationen neuesten Datums.

A. Gabert

Acta Biotechnol. 3 (1983) 3, 251–260

Enzymologie der n-Alkanoxidation bei Acinetobacter

H.-P. Kleber, R. Claus und O. Asperger

Karl-Marx-Universität Leipzig, Sektion Biowissenschaften,
Bereich Biochemie, 7010 Leipzig, Talstr. 33

Biotechnologie-Symposium 1982, Leipzig, DDR

Summary

Enzymology of n-alkane degradation was studied by using different *Acinetobacter* sp.. Cell-free extracts of *A. calcoaceticus* 69/V oxidize the n-alkane to the corresponding fatty acid via the intermediately formed n-alkanol. Soluble and particulate components are necessary for activity. Independently, a rubredoxin and a NADH-dependent rubredoxin reductase were isolated from this strain. Rubredoxin is induced by n-alkanes. The occurence of a cytochrome P-450 was observed only in other strains of *Acinetobacter*.
A. calcoaceticus 69/V contains a particulate pyridine nucleotide-independent and a soluble NADP-dependent alcohol dehydrogenase; the particulate aldehyde dehydrogenase is also NADP-dependent. The particulate dehydrogenases are inducible.
The results concerning the regulation of enzymes of the citrate and glyoxylate cycle as well as of the malic enzyme indicate that the corresponding fatty acids formed from the alkanes are metabolized via β-oxidation.

Zusammenfassung

Die Enzymologie des n-Alkanabbaus verschiedener *Acinetobacter*-Species wurde untersucht. Zellfreie Extrakte von *A. calcoaceticus* 69/V bauen n-Alkane über intermediär gebildete Alkohole zu den korrespondierenden Fettsäuren ab. Lösliche und partikuläre Komponenten sind für die Aktivität erforderlich. Daneben wurden aus diesem Stamm ein Rubredoxin und eine NADH-abhängige Rubredoxin-Reductase isoliert. Rubredoxin wird durch n-Alkane induziert. Im Gegensatz dazu wurde in anderen *Acinetobacter*-Stämmen Cytochrom P-450 nachgewiesen.
A. calcoaceticus 69/V enthält neben einer partikulären Pyridinnucleotid-unabhängigen eine lösliche NADP-abhängige Alkoholdehydrogenase; die partikuläre Aldehyddehydrogenase ist ebenfalls NADP-abhängig. Die partikulären Dehydrogenasen sind induzierbar.
Die Ergebnisse zur Regulation von Enzymen des Citrat- und Glyoxylatzyklus sowie des Malatenzyms deuten darauf hin, daß die aus den n-Alkanen gebildeten Fettsäuren durch β-Oxidation abgebaut werden.

Über den mikrobiellen Abbau aliphatischer Kohlenwasserstoffe existieren in der Literatur zahlreiche, meist ältere und aus mikrobiologischer Sicht verfaßte Übersichtsartikel [Lit. in 1], aber auch einige jüngere, in denen biochemische Aspekte im Vordergrund stehen [1–4]. Im folgenden soll eine Übersicht über die von uns an verschiedenen *Acinetobacter*-Stämmen gewonnenen Ergebnisse über den enzymatischen Abbau mittel- und langkettiger n-Alkane gegeben werden.

Tabelle 1. Verwertung von n-Alkanen unterschiedlicher Kettenlänge durch verschiedene *Acinetobacter*-Stämme bei Kultivierung auf Mineral-Agarplatten

C-Quelle	*Acinetobacter*-Stämme											
	69-V	EB[a] 114	EB 106	EB 105	EB 102	EB 104	EB 113	CCM[b] 5593	CCM 2355	CCM 5594	CCM 5595	CCM 5572	CCM 2376
n-Hexan	—	—	—	—	—	—	—	—	—	—	—	—	—
n-Heptan	—	—	—	—	—	++	++	—	—	—	—	—	—
n-Octan	n. b.	n. b.	n. b.	n. b.	n. b.	n. b.	n. b.	—	—	—	—	—	—
n-Nonan	—	—	—	(+)	(+)	+++	+++	—	—	—	—	—	—
n-Decan	—	—	—	(+)	n. b.	+++	n. b.	—	—	—	—	—	—
n-Undecan	+	+	+	(+)	(+)	+++	+++	—	+	(+)	+	(+)	(+)
n-Dodecan	+++	++	+++	++	++	+++	+++	—	+++	+++	+++	+++	+++
n-Tetradecan	+++	+++	+++	+++	+++	+++	+++	—	+++	(+)	+++	(+)	+++
n-Pentadecan	+++	+++	+++	+++	+++	+++	+++	(+)	+++	(+)	+++	(+)	(+)
n-Hexadecan	+++	+++	+++	+++	+++	+++	+++	(+)	+++	+++	+++	+++	(+)
n-Heptadecan	+++	+++	+++	+++	+++	+++	+++	(+)	+++	+++	+++	++	+++
n-Octadecan	++	++	++	n. b.	n. b.	+++	n. b.	(+)	++	++	++	++	++
Parex	+++	+++	+++	+++	+++	+++	+++	(+)	+++	+++	+++	+++	+++

—: kein Wachstum; +: schwaches Wachstum; +++: gutes Wachstum; n. b.: nicht bestimmt; (+): unsichere Beurteilung

[a] Stammsammlung des Instituts für Technische Chemie der AdW der DDR, Leipzig, DDR.
[b] Tschechoslowakische Stammsammlung für Mikroorganismen, U. E. Purkyne Universität, Brno, CSSR.

Der erste Stamm, den wir vor Jahren aus Bodenproben isoliert und untersucht haben, wurde als *A. calcoaceticus* (69/V) [5] identifiziert und verwertet langkettige n-Alkane mit mehr als 10 C-Atomen als einzige C-Quelle (Tab. 1). 1-Bromalkane werden dabei genauso gut wie die entsprechenden nichtsubstituierten Verbindungen verwertet [5]. Von den 1-Alkanolen werden auch Substrate mit kürzerer Kettenlänge metabolisiert. Alle untersuchten n-Alkansäuren und 1-ω-Alkansäuren werden — unabhängig von der Zahl der C-Atome — mit Ausnahme der C_3-Verbindungen sehr gut verwertet [5].
Das Wachstum auf n-Alkanen bedingt — wie an *A. calcoaceticus* 69/V gezeigt werden konnte — morphologische Veränderungen der Feinstruktur des Bakteriums [6]. Ähnlich wie von FINNERTY und seinen Mitarbeitern [7, 8] für *Acinetobacter* HO1-N beschrieben, sind nach Wachstum auf Hexadecan — im Gegensatz zur Verwertung von Succinat — Inklusionen zu erkennen, die, wie gaschromatographisch gezeigt werden konnte, Hexadecan enthalten. Die Hexadecan-Inklusionen sind dabei von "unit-membrane"-Aggregaten umgeben und scheinen in direkter Verbindung mit der Cytoplasmamembran zu stehen. Cytoplasmamembran und Membranaggregate haben die gleiche Dicke von 6,5 nm [6]. Die Membranhüllen sind dabei offenbar der Sitz von Enzymen, die am Kohlenwasserstoffabbau beteiligt sind. So konnte kürzlich von VOŘISEK et al. [9] nachgewiesen werden, daß die membrangebundene NADP-abhängige Aldehyddehydrogenase aus *A. calcoaceticus* an der Oberfläche von Kohlenwasserstoff-enthaltenden kleinen Vesikeln lokalisiert ist.
Auf der Grundlage von Versuchsergebnissen zur Oxidation von n-Hexadecan und dessen Derivaten nach Wachstum von *A. calcoaceticus* 69/V auf unterschiedlichen C-Quellen postulierten AURICH und EITNER [10] — ähnlich wie für andere Bakterien beschrieben — folgenden Abbauweg des n-Hexadecans:

n-Hexadecan → 1-Hexadecanol → Hexadecanal → Hexadecansäure.

Dabei konnte gezeigt werden, daß der verwendete Testorganismus zur Oxidation von n-Alkanen mindestens ein Enzym durch Induktion neu bilden muß [10]. Als Induktoren dafür können n-Alkane fungieren. Alkohol, Aldehyd und Fettsäure sind — im Gegensatz

Tabelle 2. Zellfreie Alkanoxidation bei Bakterien

Bakterienstamm	Kultur-substrat	Test-substrat	Autoren
Pseudomonas oleovorans	n-Hexan	n-Octan	BAPTIST et al. (1963) [16]
Pseudomonas denitrificans	n-Hexan	n-Decan	KUSUNOSE et al. (1967) [17]
Pseudomonas aeruginosa (Sol 20)	n-Heptan	n-Heptan	AZOULAY et al. (1963) [18]
Pseudomonas aeruginosa (P 473)	n-Heptan	n-Heptan	VAN EYK und BARTELS (1970) [19]
Pseudomonas aeruginosa ATCC 17423	n-Heptan	n-Heptan	MACHAM und HEYDEMAN (1974) [20]
Pseudomonas aeruginosa S7B1	n-Hexadecan	n-Hexadecan	MATSUYAMA et al. (1981) [15]
Pseudomonas species	n-Heptan	n-Decan	HAMMER und LIEMANN (1976) [21]
Corynebacterium sp. Stamm 7E1C	n-Octan	n-Octan	CARDINI und JURTSCHUK (1970) [22]
Mycobacterium rhodochrous; Stamm M 35	n-Nonan	n-Octan	JENKINS et al. (1972) [23]
Marines Bakterium, Stamm L. 16.1	n-Hexadecan	n-Decan	BERTRAND et al. (1976) [14]

z. B. zu den Befunden bei *Candida lipolytica* [11] — offenbar dazu nicht in der Lage. Diese — mittels der Warburg-Technik erhaltenen — Resultate konnten durch polarographische O_2-Messungen bestätigt werden [12].
Parallel zu diesen Versuchen gelang auch der Nachweis der Oxidation von n-Tetradecan zu n-Tetradecansäure durch zellfreie Extrakte von *A. calcoaceticus* [13]. Das für den Umsatz erforderliche Enzymsystem erfordert neben NADH Sauerstoff und besteht aus löslichen und partikulären Komponenten.
Wenn man die bisher publizierten Ergebnisse über die zellfreie Alkanoxidation betrachtet, so ist auffallend, daß als Substrate entweder nur Alkane bis zu 10 Kohlenstoffatomen eingesetzt wurden, oder die maximalen Umsatzraten eindeutig bei den Alkanen dieser Kettenlängen lagen (Tab. 2). *A. calcoaceticus*, ein von Bertrand et al. [14] untersuchtes marines Bakterium, sowie *Pseudomonas aeruginosa* S7B1 [15] sind nach unserer Kenntnis die einzigen Bakterienstämme, die auf langkettigen n-Alkanen wachsen und bei denen eine zellfreie n-Alkanoxidation nachgewiesen wurde. Der primäre Alkohol war dabei auch in unseren Versuchen das erste stabile Zwischenprodukt [13].
Die Schlüsselreaktion und wahrscheinlich der geschwindigkeitsbestimmende Schritt scheint dabei auch in *Acinetobacter* der Angriff des Sauerstoffs auf das n-Alkanmolekül zu sein. Für diese primäre Reaktion werden bekanntlich drei Mechanismen diskutiert, von denen der Hydroxylierungsmechanismus derzeit experimentell am besten belegt ist.

Schematische Darstellung von Hydroxylierungssystemen

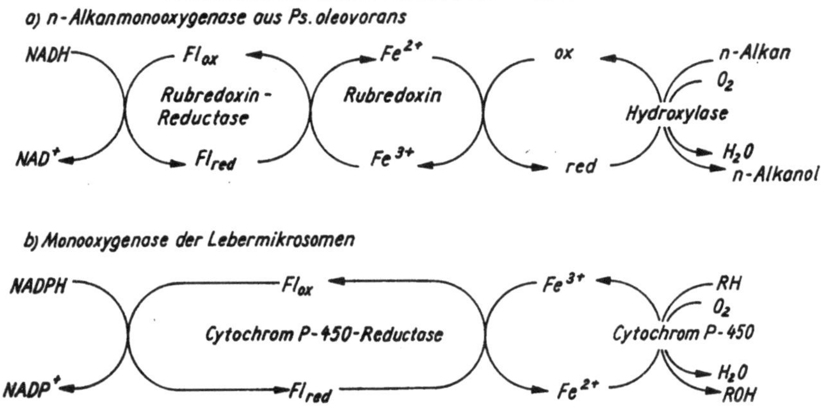

Wie die schematische Darstellung zeigt, bestehen Hydroxylierungssysteme aus einer eigentlichen Hydroxylase, die Substrat und Sauerstoff bindet und einem Elektronentransportsystem, das entweder aus einer einzelnen Reductase (soweit bekannt, membrangebunden und mit NADPH als Elektronendonator) oder aus einem niedermolekularen Protein, einem Eisen-Schwefel-Protein, und der entsprechenden Reductase (soweit bekannt, lösliche Proteine) besteht.
Bei n-alkanverwertenden Hefen [u. a. 24] und bei einem alkanassimilierenden Corynebacterium [22] wurde Cytochrom P-450 als terminale Oxidase gefunden, während bei *Pseudomonas oleovorans* [25] Rubredoxin als Elektronentransportkomponente und ein nicht exakt charakterisiertes Eisen-haltiges Protein als terminale Oxidase nachgewiesen wurden.
Auf der Suche nach Komponenten des postulierten n-Alkanhydroxylierungssystems untersuchten wir die Cytochromzusammensetzung von *A. calcoaceticus* nach Wachstum auf verschiedenen C-Quellen [26]. Dabei zeigte sich, daß *A. calcoaceticus* — in Übereinstimmung mit den Ergebnissen bei *A. anitratus* [27] sowie *Acinetobacter* HO1-N [28] —

die Cytochrome b, o und d enthält. Die Abwesenheit von Cytochrom c steht ebenfalls in Einklang mit den Resultaten anderer Autoren. Bei der Kultivierung auf den verschiedenen C-Quellen konnten wir — vom Fehlen des Cytochrom d nach Wachstum auf Hexadecan abgesehen, welches aber auch nach Wachstum auf anderen C-Quellen nur in geringer Konzentration vorhanden war — keine signifikanten Unterschiede in der Cytochromzusammensetzung finden. Nach Wachstum auf Hexadecan wurde weder Cytochrom P-450 noch eine andere als die schon beschriebenen terminalen Oxidasen gefunden. Da das Verhältnis zwischen terminalen und nichtterminalen Oxidasen bei *A. calcoaceticus* 69/V weitgehend konstant bleibt, auch wenn die C-Quelle variiert wird [26], ist es unwahrscheinlich, daß ein Cytochrom zur Hydroxylierung von n-Alkanen induziert wird. Die Hydroxylierung der n-Alkane sollte deshalb bei *A. calcoaceticus* 69/V über einen anderen Mechanismus erfolgen. Dabei könnte das in diesem Stamm nachgewiesene Rubredoxin eine Rolle spielen [29].
Einige wichtige Eigenschaften des isolierten Rubredoxins sind in Tab. 3 zusammengefaßt. Sie schließen das Vorliegen eines Ferredoxins aus.

Tabelle 3. Eigenschaften von Rubredoxin aus *Acinetobacter calcoaceticus* (1500fach angereichert; diskelektrophoretisch rein)

Molekulargewicht	6000 bzw. 12000
Aminosäurezusammensetzung	11 saure und 6 basische Aminosäuren
Säurelabiler Schwefel	nicht nachweisbar
Polypeptidkette	4 Cysteinreste und 1 Atom Fe
Absorptionsmaxima	λ 280, 380 und 490 nm

Während uns bisher die Isolierung der n-Alkanhydroxylase aus *A. calcoaceticus* noch nicht gelang, konnten wir — ähnlich wie in *Ps. oleovorans* [30] — in unserem Testorganismus neben dem Rubredoxin eine Rubredoxin-Reductase nachweisen und isolieren [31].
Die aus unserem Teststamm nach Wachstum auf n-Alkanen angereicherte Rubredoxin-Reductase katalysiert spezifisch den Elektronentransport von NADH über Rubredoxin — nicht aber über Ferredoxin aus Spinat oder Adrenodoxin — zum Cytochrom c [32]. Das Enzym entspricht in den wesentlichen kinetischen Eigenschaften der Rubredoxin-Reductase aus *Ps. oleovorans* [30].
Da sowohl in dem relativ gut untersuchten Enzymsystem aus *Ps. oleovorans* und natürlich auch in *Acinetobacter* nicht vollständig gesichert ist, ob die isolierten Einzelkomponenten Rubredoxin und Rubredoxin-Reductase die in vivo tatsächlich wirksamen Elektronendonator bzw. -übertragungssysteme sind, versuchten wir über Regulationsexperimente Aufschluß darüber zu erhalten.
Die für einen Induktionsnachweis erforderliche quantitative Bestimmung dieser Proteine in ganzen Zellen oder zellfreien Extrakten stößt auf wesentliche experimentelle Schwierigkeiten. Wir führten solche Untersuchungen für Rubredoxin aus *A. calcoaceticus* deshalb mit Hilfe eines Radioimmuntestes durch [35], nachdem wir Rubredoxin aus diesem Stamm rein dargestellt hatten und ein entsprechendes Antiserum gewinnen konnten. Mit Hilfe dieser Methode wurde eine deutliche Zunahme des Rubredoxingehaltes der Zellen während des Wachstums auf Hexadecan beobachtet (Abb. 1). Bei Kultivierung auf anderen C-Quellen war Rubredoxin nach dieser Methode in keiner Wachstumsphase nachzuweisen [34]. Dieses Ergebnis spricht zumindest für eine Beteiligung des Rubredoxins am n-Alkanstoffwechsel. Gleichzeitig wurde damit erstmalig der Nachweis der Induzierbarkeit eines Rubredoxins durch n-Alkanwachstum in Bakterien erbracht. Im Gegensatz zum Rubredoxin ist die Rubredoxin-Reductase in *A. calcoaceticus* konstitutiv [35].

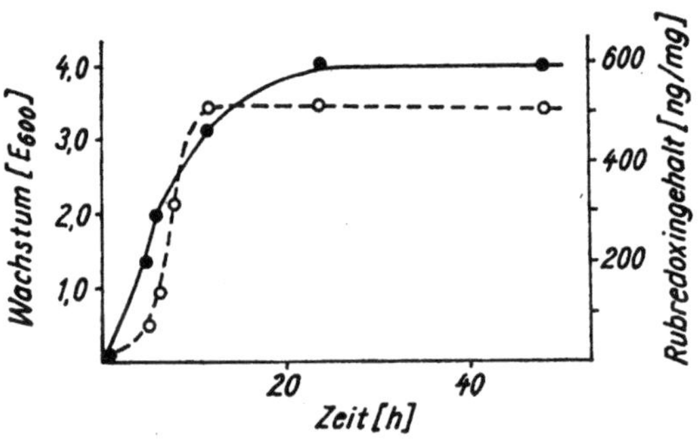

Abb. 1. Rubredoxingehalt in *Acinetobacter calcoaceticus* (69/V) (●—●, ng/mg Protein) in Abhängigkeit von der Wachstumsphase (○—○, E_{600}) mit Hexadecan (6 g/l) als C-Quelle

Tabelle 4. Cytochromzusammensetzung einiger *Acinetobacter*-Stämme nach Wachstum auf Hexadecan

Stamm	Cytochrom a	Cytochrom b	Cytochrom c	Cytochrom d	Cytochrom o	Cytochrom P-450
A. calcoaceticus 69/V[a]	–	+	–	–	+	–
A. calcoaceticus CCM 2355[b]	–	+	–	–	+	–
A. calcoaceticus CCM 5594	–	+	–	–	+	–
A. calcoaceticus CCM 5595	–	+	–	–	+	–
Acinetobacter-Species EB 114[c]	–	+	–	–	+	–
Acinetobacter-Species EB 106	–	+	–	–	+	–
Acinetobacter-Species EB 102	–	+	–	–	+	+
Acinetobacter-Species EB 103	–	+	–	–	+	+
Acinetobacter-Species EB 104	–	+	–	–	+	+
Acinetobacter-Species EB 113	–	+	–	–	+	+
Acinetobacter-Species EB 10 C	–	+	–	–	+	+
Acinetobacter-Species EB 10 D	–	+	–	–	+	+
Acinetobacter-Species EB 11	–	+	–	–	+	+
A. lwoffi CCM 5572	–	+	–	–	+	–
A. lwoffi CCM 2376	–	+	–	–	+	–

[a] Stammsammlung des Bereiches Biochemie der Sektion Biowissenschaften, Karl-Marx-Universität Leipzig, DDR
[b] Tschechoslowakische Stammsammlung für Mikroorganismen, U. E. Purkyne Universität, Brno, CSSR
Stammsammlung des Institutes für technische Chemie der AdW der DDR, Leipzig, DDR

Da uns trotz der Isolierung von Rubredoxin und der Rubredoxin-Reductase eine Identifizierung der eigentlichen Hydroxylase nicht gelang, versuchten wir durch den Nachweis von Cytochrom P-450, welches von Cardini und Jurtschuk [22] als terminale Oxidase eines Alkanhydroxylierungssystems in Corynebakterien postuliert wurde, in anderen *Acinetobacter*-Species dieser Frage etwas näherzukommen. Die Mehrzahl der untersuchten *Acinetobacter*-Stämme wuchs gut auf Hexadecan und anderen langkettigen Kohlenwasserstoffen als einziger C- und Energie-Quelle (vgl. Tab. 1). Erwähnenswert erscheint, daß zwei Stämme (EB 104 und 113) bei Angebot über die Gasphase auch Heptan gut verwerten. Die Cytochromzusammensetzung entspricht im wesentlichen der von *A. calcoaceticus* 69/V (Tab. 4). Im Gegensatz zu diesem Stamm und den CCM-Stämmen zeigen die EB-Stämme neben einem Cytochrom o-Maximum bei 417 nm einen Peak bei 450 nm im „reduziert + CO minus reduziert" Differenzspektrum [36].
Nach Wachstum von *Acinetobacter* EB 104 auf Nicht-Kohlenwasserstoff-Substraten war der Peak bei 449 nm nicht im CO-Differenzspektrum nachweisbar. Hingegen wurde Cytochrom o gefunden (Tab. 5). Damit gehören diese *Acinetobacter*-Stämme neben den Corynebakterien zu den einzigen Bakterien, bei denen nach Wachstum auf n-Alkanen Cytochrom P-450 nachgewiesen werden konnte.

Tabelle 5. Terminale Oxidasen in *A. calcoaceticus* (Stamm EB 104) nach Wachstum auf verschiedenen C-Quellen

C-Quelle	Cytochrom o	Cytochrom P-450
Hexadecan	+	+
Succinat	+	–
Succinat + Hexadecan	+	+
Hefeextrakt	+	–
Hefeextrakt + Hexadecan	+	+
Nährbouillon	+	–
Nährbouillon + Hexadecan	+	+
Hexadecanol	+	–
Palmitinsäure	+	–

Die in Tab. 5 aufgeführten Ergebnisse deuten darauf hin, daß Cytochrom P-450 durch n-Alkane induziert wird. Eine Derepression der Synthese kann ausgeschlossen werden. Eine Induktion durch n-Alkanoxidationsprodukte erscheint unwahrscheinlich, da Palmitinsäure zwar verwertet wird, aber nicht die Induktion von Cytochrom P-450 auslöst. Entsprechendes gilt für n-Hexadecanol-(1).
Eine ω-Hydroxylierung der Fettsäuren, wie sie für *Bacillus megaterium* [37] beschrieben wurde, scheint für *Acinetobacter* unwahrscheinlich.
Die Induktion von Cytochrom P-450 durch Zugabe von Hexadecan zu logarithmisch auf Nährbouillon wachsenden Zellen von *Acinetobacter*-Species EB 104 ist in Abb. 2 dargestellt. Eine Cytochrom P-450-Induktion konnte bei Zugabe von Kohlenwasserstoffen der Kettenlänge C_5—C_{20} nachgewiesen werden, wobei — in Abhängigkeit von den Kulturbedingungen — Cytochrom P-450-Konzentrationen erhalten wurden, die mit denen der Mikrosomen von Alkan-verwertenden Hefen [38] sowie von *Ps. putida* nach Wachstum auf Campher [39] vergleichbar waren.
Das Cytochrom P-450 aus *A. calcoaceticus* EB 104 konnte in Gegenwart von Triton X-100 elektrophoretisch homogen isoliert werden und wird durch ein NADH-abhängiges Ferredoxin-Reductase/Ferredoxin-System, das aus dem gleichen Stamm angereichert wurde, enzymatisch reduziert [40].
An der Weitermetabolisierung des Alkohols zur korrespondierenden Fettsäure sind Dehydrogenasen beteiligt, die sich in der Coenzymabhängigkeit und Lokalisation, d. h.

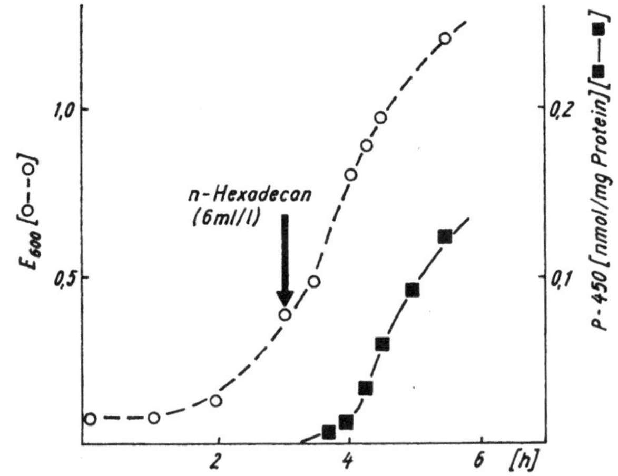

Abb. 2. Induktion von Cytochrom P-450 in *Acinetobacter*-Species EB 104 durch Hexadecan. Der Zusatz von Hexadecan (6 ml/l) erfolgte zu dem durch den Pfeil gekennzeichneten Zeitpunkt

löslich oder membrangebunden, unterscheiden. Für *A. calcoaceticus* konnten neben einer löslichen, NADP-abhängigen [41] eine partikuläre, Pyridinnucleotid-unabhängige Alkoholdehydrogenase [42] sowie eine partikuläre, NADP-abhängige Aldehyddehydrogenase [43] nachgewiesen werden. Während die NADP-abhängige Alkoholdehydrogenase konstitutiv ist [41], sind die Pyridinnucleotid-unabhängige Alkoholdehydrogenase und die NADP-abhängige Aldehyddehydrogenase induzierbar. Letztere wird dabei durch n-Alkane und Alkanole induziert, wobei die Induktionswirkung praktisch erst ab einer Kettenlänge von C_6 eintritt [44]. Die Versuchsergebnisse deuten darauf hin, daß in *A. calcoaceticus* — ähnlich wie in *Ps. putida* [45, 46] — die Pyridinnucleotid-unabhängige Alkoholdehydrogenase für die n-Alkanassimilation von größerer Bedeutung ist als die NAD- bzw. NADP-abhängigen Alkoholdehydrogenasen.

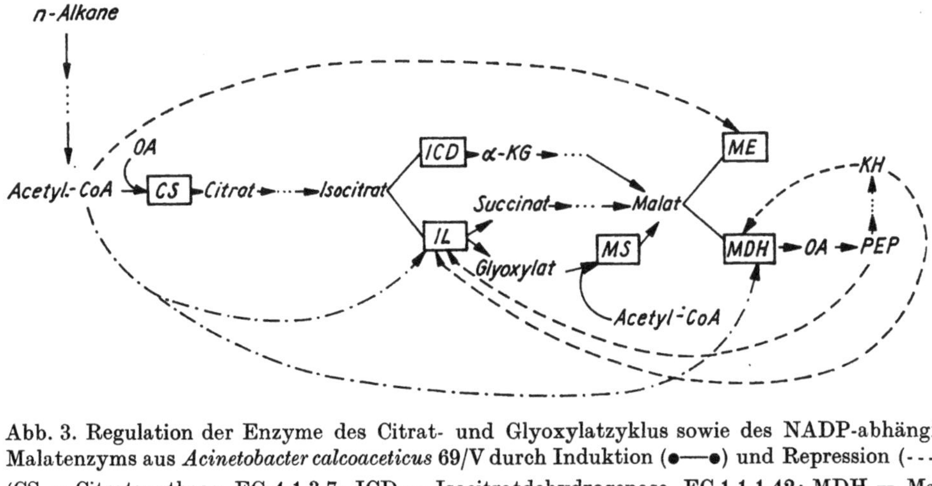

Abb. 3. Regulation der Enzyme des Citrat- und Glyoxylatzyklus sowie des NADP-abhängigen Malatenzyms aus *Acinetobacter calcoaceticus* 69/V durch Induktion (●—●) und Repression (- - -).
(CS = Citratsynthase, EC 4.1.3.7; ICD = Isocitratdehydrogenase, EC 1.1.1.42; MDH = Malatdehydrogenase, EC 1.1.1.37; IL = Isocitratlyase, EC 4.1.3.1; MS = Malatsynthase, EC 4.1.3.2; ME = Malatenzym, EC 1.1.1.40)

In Abb. 3 sind die Ergebnisse über die Regulation von Enzymen des Citrat- und Glyoxylatzyklus sowie des Malatenzyms aus *A. calcoaceticus* durch Induktion und Repression schematisch dargestellt. Durch n-Alkane bzw. Acetat werden die Isocitratlyase [47] und Malatdehydrogenase [48] induziert, das Malatenzym hingegen reprimiert [49]. Glucose und Phosphoenolpyruvat reprimieren die Isocitratlyase [47]. Um den Verlust von Säuren des Citratzyklus für biosynthetische Prozesse auszugleichen, ist beim Wachstum von Mikroorganismen auf Acetat — aber auch auf n-Alkanen — eine kontinuierliche Synthese von C_4-Dicarbonsäuren erforderlich. Diese Kompensation erfolgt im allgemeinen durch die Tätigkeit des Glyoxylatzyklus. Die Repression des Malatenzyms durch Acetat gliedert sich dabei sinnvoll in dieses Regulationsgeschehen ein, da somit ein Abfluß von Malat aus dem Citratzyklus verhindert wird. Dieser Vorgang wird noch dadurch unterstützt, daß neben Fettsäuren [50] Acetyl CoA — ähnlich wie in *E. coli* [51] und *Paracoccus denitrificans* [52] — als Endprodukt der Reaktionskette hemmend auf das Malatenzym aus *A. calcoaceticus* wirkt [53]. Insgesamt sprechen diese Ergebnisse dafür, daß zumindest ein Teil der beim n-Alkanabbau entstandenen Fettsäuren durch β-Oxidation abgebaut werden.

Literatur

[1] Aurich, H.: Sitzungsberichte der AdW der DDR. Nr. 16/N, Akademie-Verlag, Berlin, 1979.
[2] Finnerty, W. R.: Trends Biochem. Sci. **2** (1977) 73.
[3] Ratledge, C.: In: Developm. Biodegrad. Hydrocarb., p. 1. Appl. Sci. Publ., Hull, 1977.
[4] Rehm, H. J., Reiff, J.: Adv. Biochem. Engng. **19** (1981) 175.
[5] Kleber, H.-P., Schöpp, W., Aurich, H.: Z. Allg. Mikrobiol. **13** (1973) 445.
[6] Müller, H., Naumann, A., Claus, R., Kleber, H.-P.: Z. Allg. Mikrobiol. (im Druck).
[7] Kennedy, R. S., Finnerty, W. R., Sudarsanan, K., Young, R. A.: Arch. Microbiol. **102** (1975) 75.
[8] Scott, C. C. L., Finnerty, W. R.: J. Bacteriol. **127** (1976) 481.
[9] Vořisek, J., Sorger, H., Aurich, H., Lojda, Z.: Sympos. histochem. Karlovy Vary (Abstract) 1982.
[10] Aurich, H., Eitner, G.: Z. Allg. Mikrobiol. **13** (1973) 539.
[11] Nyns, E. J., Auqiere, J. P., Wiaux, A. L.: Z. Allg. Mikrobiol. **9** (1969) 373.
[12] Asperger, O., Aurich, H.: Z. Allg. Mikrobiol. **17** (1977) 419.
[13] Asperger, O., Futtig, A., Behrends, B., Kleber, H.-P. , Aurich, H.: Wiss. Z. Karl-Marx-Univ. Leipzig, Math.-Nat. R. **27** (1978) 3.
[14] Bertrand, J. D., Doux, H. J. M., Azoulay, E.: Biochimie **58** (1976) 843.
[15] Matsuyama, H., Nakahara, T., Minoda, Y.: Agric. Biol. Chem. **45** (1981) 9.
[16] Baptist, J. N., Gholson, R. K., Coon, J. M.: Biochim. biophys. Acta **69** (1963) 40.
[17] Kusunose, M., Matsumoto, J., Ichihara, K., Kusunose, E., Nozaka, J.: J. Biochem. **61** (1967) 665.
[18] Azoulay, E., Chouteau, J., Davidovics, G.: Biochim. biophys. Acta **77** (1963) 554.
[19] van Eyk, J., Bartels, T. J.: J. Bacteriol. **104** (1970) 1065.
[20] Macham, L. P., Heydeman, M. T.: J. gen. Microbiol. **85** (1974) 77.
[21] Hammer, K. D., Liemann, F.: Z. Bakt. Hyg., I. Abt. Orig. B **162** (1976) 169.
[22] Cardini, G., Jurtschuk, P.: J. Biol. Chem. **245** (1970) 2789.
[23] Jenkins, P. G., Raboin, D., Moran, F.: J. gen. Microbiol. **72** (1972) 395.
[24] Gallo, M., Bertrand, J. C., Azoulay, E.: FEBS Lett. **19** (1971) 45.
[25] Coon, M. J., Autor, A. P., Boyer, R. F., Lode, E. T., Strobel, H. W.: In: King, T. E., Mason, M. S. and Morrison, M. (eds.): Oxidases and related redox systems. University Park Press Baltimore, London, Tokyo, 1973, p. 529.
[26] Asperger, O., Kleber, H.-P., Aurich, H.: Acta biol. med. germ. **37** (1978) 191.
[27] Spicher, G.: Z. Bakt. Hyg., I. Abt. Orig. A **228** (1974) 317.
[28] Ensley, B. D., Finnerty, W. R.: J. Bacteriol. **142** (1980) 859.
[29] Aurich, H., Sorger, D., Asperger, O.: Acta biol. med. germ. **35** (1976) 443.

[30] Ueda, T., Coon, M. J.: J. Biol. Chem. **247** (1972) 5010.
[31] Claus, R., Asperger, O., Kleber, H.-P.: Wiss. Z. Karl-Marx-Univ. Leipzig, Math.-Nat. R. **27** (1978) 17.
[32] Claus, R., Asperger, O., Kleber, H.-P.: Z. Allg. Mikrobiol. **19** (1979) 695.
[33] Claus, R., Hädge, D., Asperger, O., Fiebig, H., Kleber, H.-P.: Z. Allg. Mikrobiol. **20** (1980) 95.
[34] Claus, R., Asperger, O., Kleber, H.-P.: Arch. Microbiol. **128** (1980) 263.
[35] Claus, R., Kleber, H.-P.: Z. Allg. Mikrobiol. **22** (1982) 3.
[36] Asperger, O., Naumann, A., Kleber, H.-P.: FEMS Microbiol. Lett. **11** (1981) 309.
[37] Matson, R. S., Hare, R. S., Fulco, A. J.: Biochim. biophys. Acta **487** (1977) 487.
[38] Müller, H.-G., Schunck, W.-H., Riege, P., Honeck, H.: Acta biol. med. germ. **38** (1979) 345.
[39] Peterson, J. A.: J. Bacteriol. **103** (1970) 714.
[40] Asperger, O., Müller, R., Kleber, H.-P.: Acta Biotechnol. (im Druck).
[41] Eitner, G.: Dissertation zur Promotion A. Karl-Marx-Universität Leipzig 1973.
[42] Tauchert, H., Roy, M., Schöpp, W., Aurich, H.: Z. Allg. Mikrobiol. **15** (1975) 457.
[43] Sorger, H., Aurich, H.: Wiss. Z. Karl-Marx-Univ. Leipzig, Math.-Nat. R. **27** (1978) 35.
[44] Aurich, H., Eitner, G.: Z. Allg. Mikrobiol. **17** (1977) 263.
[45] Grund, A., Shapiro, J., Fennewald, M., Bacha, P., Leahy, J., Markbreiter, K., Nieder, M., Toepfer, M.: J. Bacteriol. **123** (1975) 546.
[46] Aurich, H., Tauchert, H., Grunow, M.: 10. FEBS-MEETING, Paris 1975.
[47] Kleber, H.-P., Aurich, H.: Z. Allg. Mikrobiol. **13** (1973) 473.
[48] Kleber, H.-P., Aurich, H.: Z. Allg. Mikrobiol. **14** (1974) 575.
[49] Kleber, H.-P.: Z. Allg. Mikrobiol. **13** (1973) 467.
[50] Kleber, H.-P., Göbel,, H.: Z. Allg. Mikrobiol. **15** (1975) 75.
[51] Sanwal, B. D., Wright, J. A., Smando, R.: Biochem. biophys. Res. Commun. **31** (1968) 623.
[52] Slabas, A. R., Whatley, F. R.: Biochem. biophys. Res. Commun. **74** (1977) 1116.
[53] Kleber, H.-P.: Z. Allg. Mikrobiol. **15** (1975) 19.

Acta Biotechnol. 3 (1983) 3, 261–268

Methanol Oxidation and Growth Yields in Methylotrophic Bacteria: A Review

C. Anthony

Biochemistry Department, University of Southampton, Southampton SO 9 3TU. U. K.

Symposium on Biotechnology 1982, Leipzig, G.D.R.

Summary

In all bacteria growing on methane or methanol the methanol is oxidised by way of methanol dehydrogenase. This unusual quinoprotein interacts with the electron transport chain at the level of cytochrome *c*. All methylotrophs having methanol dehydrogenase contain 2 different types of soluble cytochrome *c*. One of these, but not the other, reacts with methanol dehydrogenase; no intermediate electron carrier is required for this reaction. The P/O ratio for methanol oxidation is 1 although the midpoint redox potential of the methanol/formaldehyde couple suggests that a higher ATP yield might be theoretically possible. Curves are presented that allow the prediction of cell yields, and the effects of altering the system for methanol oxidation, on these cell yields.

Zusammenfassung

Methanol wird in allen auf Methan oder Methanol wachsenden Bakterien durch die Methanol-dehydrogenase oxidiert. Dieses NAD-unabhängige Chinoprotein steht mit der Atmungskette auf der Stufe des Cytochrom *c* in engem Kontakt. Alle Methanoldehydrogenase besitzenden methylo-otrophen Mikroorganismen enthalten zwei unterschiedliche Typen von löslichem Cytochrom *c*. Nur einer davon reagiert direkt ohne zwischengeschalteten Elektronencarrier mit der Methanol-dehydrogenase. Der P/O-Quotient für die Methanoloxidation beträgt 1, obwohl das Normal-Redoxpotential für die Umwandlung von Methanol zu Formaldehyd theoretisch eine höhere ATP-Ausbeute ermöglichen würde. Es werden Kurven vorgelegt, die eine Vorhersage der Zellausbeuten und der Effekte bei Veränderung des Methanoloxidationssystems auf diese Zellausbeuten erlauben.

All bacteria growing on methane or methanol oxidise methanol by way of methanol dehydrogenase which is an unusual, NAD-independent dehydrogenase having pyrrolo-quinoline quinone (PQQ) as its prosthetic group [1, 2]. Because cell carbon is assimilated at the level of formaldehyde, every molecule of methanol metabolised by the bacteria must be oxidised by this dehydrogenase. The midpoint redox potential of the HCHO/CH_3OH couple is −0.182 V; it might therefore be expected that methanol dehydrogenase would interact with the electron transport chain at or before the level of cytochrome *b*, and that the P/O ratio would be 2 for the oxidation of methanol to formaldehyde. However, all the available evidence indicates that methanol dehydrogenase interacts at the level of cytochrome *c*, thus by-passing cytochrome *b* and the earlier energy-transducing parts of the electron transport chain [3, 2]. This is consistent with

the fact that the P/O ratio for oxidation of methanol to formaldehyde is only 1, and not 2 as might have been expected.
This raises a number of questions with respect to electron transport and energy transduction during methanol oxidation, which this review will briefly attempt to answer.

1. How many cytochromes *c* are present in methylotrophs?
2. Is the reaction between methanol dehydrogenase and cytochrome *c* direct, or is there an intermediate electron carrier?
3. Is there a special cytochrome *c* involved in methanol oxidation?
4. How are methanol dehydrogenase and cytochrome *c* arranged to generate ATP?
5. Does the unusual system for methanol oxidation have any special effect on growth yields of methylotrophs, and could a genetically-engineered change in this system for methanol oxidation lead to greater cell yields?

1. How many cytochromes c are present in methylotrophs?

All methylotrophs studied in any detail have at least 2 soluble cytochromes *c*, and some membrane-bound cytochrome *c* which may or may not be the same as the soluble cytochromes. Some of these are listed in Table 1. We have also confirmed the presence of 2 soluble cytochromes *c* in the methane-utilising bacteria, in the facultative autotroph and methylotroph *Paracoccus denitrificans*, and in *Hyphomicrobia*. The two types of cytochrome *c* differ in their midpoint redox potentials, their molecular weights and their isoelectric points. They are completely different proteins; one type is not a dimer or degradation product of the other [7].

Table 1. Properties of the two types of soluble cytochromes *c* of methylotrophs
The data for *Pseudomonas* AM1 are from O'KEEFFE and ANTHONY [4]; for *Methylophilus methylotrophus* from CROSS and ANTHONY [5]; and for *Methylomonas* J from OHTA and TOBARI [6].

	Isoelectric point	Molecular weight	Midpoint redox potential
Pseudomonas AM 1			
Cytochrome c_H	8.8	11,000	+294 mV
Cytochrome c_L	4.2	20,900	+256 mV
Methylophilus methylotrophus			
Cyrochrome c_H	8.9	8,500	+373 mV
Cytochrome c_L	4.0–4.4	21,000	+310 mV
Methylomonas J			
Cytochrome c_{551} (I)	5.3	16,000	+290 mV
Cytochrome c_{551} (II)	4.3	12,500	+240 mV

2. Is the reaction between methanol dehydrogenase and cytochrome c direct, or is there an intermediate electron carrier?

The answer to this question has been difficult to determine unequivocally with purified proteins. This is because addition of methanol dehydrogenase to cytochrome *c* leads to reduction of the cytochrome even in the absence of methanol [8]. A less direct approach has now solved this problem. This has been achieved by using small amounts of the cytochrome *c* from methylotrophs to mediate between methanol dehydrogenase and large quantities of mammalian cytochrome *c* — used as a terminal electron acceptor [9].

Methanol-dependent reduction of the cytochrome *c* was recorded with comcomitant production of stoichiometric amounts of formaldehyde. Thus, using this system, it has now been conclusively shown that no intermediate electron carrier is required between methanol dehydrogenase and pure cytochrome *c* from methylotrophs. It should be noted that the rate measured for the methanol: cytochrome *c* oxidoreductase activity of methanol dehydrogenase was too low to account for the rate of respiration observed in whole cells; this may be related to the change in environment occurring on release of the cytochrome *c* and methanol dehydrogenase from the membrane surface of the bacteria.

3. Is there a special cytochrome *c* involved in methanol oxidation?

The fact that methylotrophs have two very distinct types of cytochrome *c* suggests that one of them may have a special role in methanol oxidation, and the results in Table 2 appear to confirm that this is the case. The first point to note from Table 2 is that neither horse heart cytochrome *c* nor the cytochrome from the non-methylotroph, *Pseudomonas aeruginosa,* is able to react with any of the 3 dehydrogenases. The second point is that only one of the two cytochromes from each methylotroph is able to react with methanol dehydrogenase. The third point is that the cytochromes *c* from *Methylophilus methylotrophus* and from *Paracoccus denitrificans* are only able to react with their own dehydrogenase, whereas the cytochrome *c* from *Pseudomonas* AM 1 is less specific and is able to react with all three dehydrogenases.

Table 2. The specificity of reaction between methanol dehydrogenases and cytocromes *c* This Table is taken from BEARDMORE-GRAY et al. [9].

Cytochrome *c*	Reduction of cytochromes by methanol dehydrogenase from the following bacteria		
	M. methylotrophus	*Pa. denitrificans*	*Pseudomonas* AM 1
Horse heart			
cytochrome *c* (pI, 10.4)	–	–	–
Pseudomonas aeruginosa			
cytochrome c_{551} (pI, 4.7)	–	–	–
M. methylotrophus			
cytochrome c_H (pI, 8.85)	–	–	–
cytochrome c_L (pI, 4.35)	+	–	–
Pa. denitrificans			
cytochrome *c* (1), (pI, acidic)	–	–	–
cytochrome *c* (2) (pI, more acidic)	–	+	–
Pseudomonas AM 1			
cytochrome c_H (pI, 8.8)	–	–	–
cytochrome c_L (pI, 4.2)	+	+	+

4. How are methanol dehydrogenase and cytochrome *c* arranged to generate ATP?

The two main options for arrangement of the 'methanol oxidase' system are shown in Fig.1. The protonmotive redox 'loop' (Fig. 1a) is unlikely because the dehydrogenase and its prosthetic group are very hydrophilic and unlikely to span a lipoprotein mem-

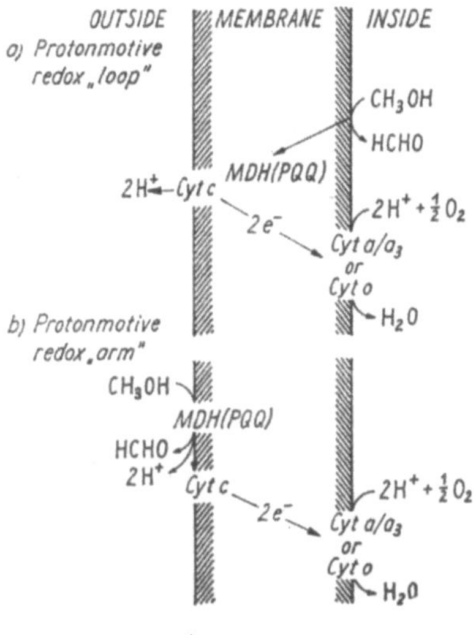

Fig. 1. Possible arrangements of methanol dehydrogenase and cytochrome *c* in the bacterial membrane. In both schemes $2H^+$ are effectively moved from the inside to the outside for every molecule of methanol oxidised to formaldehyde. In (a) the MDH is transmembranous and in (b) it is on the outer surface. [2]

brane. All the evidence at present supports the protonmotive redox 'arm' is shown in Fig. 1b [10, 11, 12, 13, 14]. This arrangement of the 'methanol oxidase' system leads to translocation of one pair of protons across the bacterial membrane and thus to a P/O ratio of 1, and this value has been confirmed by more direct measurements [15]. It is of interest that if the methanol dehydrogenase is on the outer surface, as indicated here, the formaldehyde produced must be able to cross the membrane for its further oxidation and assimilation into cell material. It will thus be of interest to determine how it is transported, and whether formaldehyde transport is involved in the regulation of oxidation and assimilation of this critical 'branch point' intermediate.

5. Does the unusual system for methanol oxidation have any special effects on cell yields of methylotrophs, and could a genetically-engineered change in this system lead to greater cell yields?

The summary above shows that there are two unusual features of the system for methanol oxidation in bacteria. The first is that no NADH is produced during the first step in the oxidation of methanol; and the second is that the P/O ratio for this step is only 1, whereas the value of 2 is theoretically possible. Table 3 illustrates one effect of these features on methylotrophic metabolism; that is, during growth on methane or methanol, usually less than 30% of electron transport is by way of NADH dehydrogenase. In methylotrophs growing on methane or methanol most electron transport is by way of the methanol dehydrogenase whose activity is associated with a P/O ratio of only 1. This raises the question of whether or not cell yields might be increased if the methanol dehydrogenase were to be changed to an NADH-yielding enzyme, or if the P/O ratio for methanol oxidation could be increased to 2.

For some methylotrophs the answer is immediately obvious; in those methylotrophs that are NADH-limited rather than ATP-limited (this includes most methanotrophs), increasing the P/O ratio for methanol oxidation would have a negligible effect on cell yields [16, 17, 2]. In these bacteria, however, if the methanol dehydrogenase were to become NADH-yielding, then the cell yields would be much higher even if the P/O ratio were to remain low.

In those methanol-utilizers having the RuMP assimilation pathway, higher yields would be obtained if the P/O ratio were to be increased or if the methanol dehydrogenase were

Table 3. Electron flow from each dehydrogenase expressed as a proportion of total electron transport

Growth substrate	Percentage of electron flow from each dehydrogenase			
	Methanol	Formaldehyde	Flavoprotein	NADH
Methane				
Serine pathway bacteria (e.g. *Methylosinus trichosporium*)	78–88	(NAD^+-linked)	9–12	0–13
RuMP pathway bacteria (e.g. *Methylococcus capsulatus*)	84–92	(NAD^+-linked)	0	8–16
Methanol				
Serine pathway bacteria				
(a) HCHO oxidation yields 1 NADH (e.g. *Pseudomonas* AM 1)	53–56	35	10	0–3
(b) HCHO oxidation yields 2 NADH	53–64	(NAD^+-linked)	9–15	21–40
RuMP pathway bacteria (e.g. *Methylophilus methylotrophus*)	57–70	(NAD^+-linked)	0	30–43
RuBP pathway bacteria (e.g. *Paracoccus denitrificans*)	52–62	(NAD^+-linked)	0	38–48
Conventional substrates (including formate)	0	0	0–26	74–100

The range of values predicted is for a range of assumed P/O ratios; the lowest is assumed to be one for each oxidation step and the highest P/O ratios are assumed to be 1, 2 or 3 for methanol dehydrogenase, flavoprotein and NADH, respectively. Higher P/O ratios lead to a lower proportion of electron transport from NADH dehydrogenase and a higher proportion from other dehydrogenases. During growth on methane only 25–40% of the total oxygen consumed is by way of electron transport and oxidases, the remainder being used in the initial hydroxylation reaction (assumed here to be NADH-linked). In serine pathway bacteria the oxidation of acetyl-CoA to glyoxylate is assumed to involve a flavoprotein.
This Table is taken from Anthony [2].

to be NADH-yielding. The extent of the expected increases can be estimated from the Fig. 2 and 3. These data have been calculated using methods previously described [16, 17, 2]. The formula for calculation of yields is:

$$12CH_3OH + 1.5NADH + 3NH_3 + x\ ATP \rightarrow (C_4H_8O_2N)_3 + 12PQQH_2$$

To use this formula for predicting yields a value for the ATP requirement for biosynthesis (x) must be chosen and values for the P/O ratios assumed for oxidation of NADH and $PQQH_2$ (reduced methanol dehydrogenase). The value for x can be calculated, or the measured yield (Carbon Conversion Efficiency) can be used for estimating a value for x for a given organism from curve A in Figs. 2 and 3. Using this same value for x the effect of changing the nature of the system for oxidation of methanol can be calculated by using the alternative curves in Figs. 2 and 3.

Curve A is the normal situation; the methanol dehydrogenase (MDH) is a PQQ-containing enzyme coupling to the electron transport chain at the level of cytochrome *c*. The P/O ratio is 1 for $PQQH_2$ and 3 for NADH oxidation.
Curve B is the same as A except that the P/O ratio for NADH is 2.
Curve C is the result of coupling the MDH so that the P/O ratio for $PQQH_2$ is 2 instead of 1 (P/O ratio for NADH is 3).

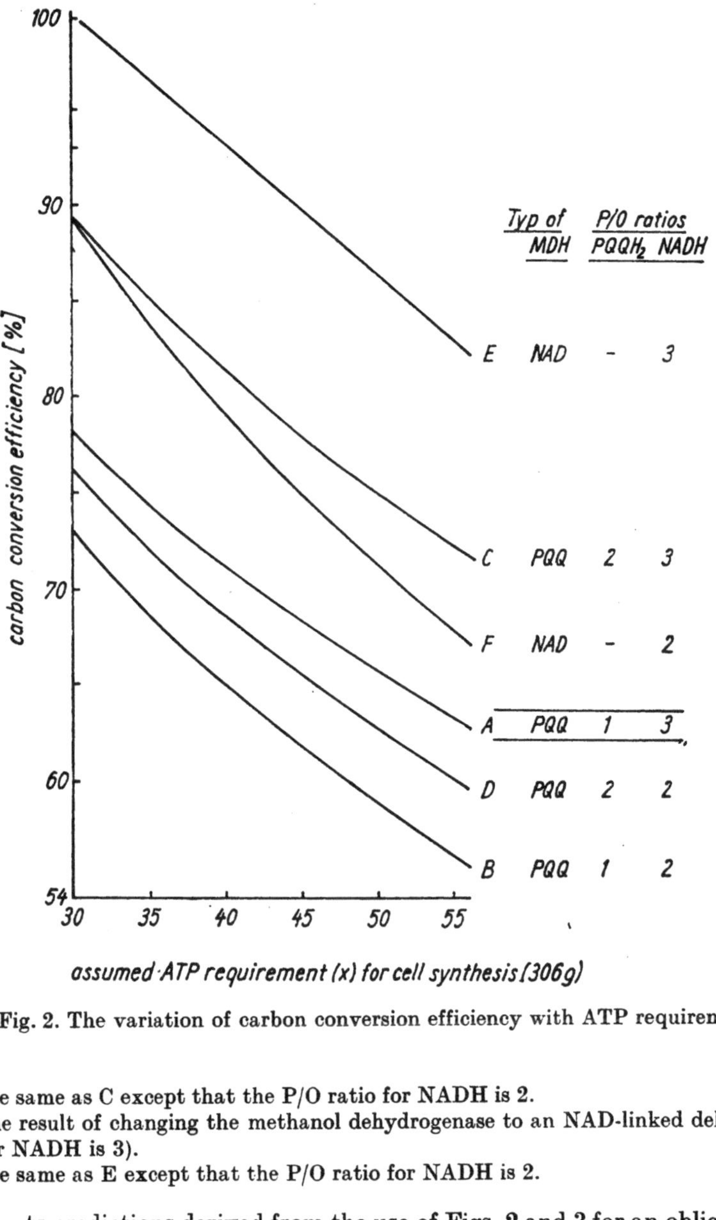

Fig. 2. The variation of carbon conversion efficiency with ATP requirement

Curve D is the same as C except that the P/O ratio for NADH is 2.
Curve E is the result of changing the methanol dehydrogenase to an NAD-linked dehydrogenase (P/O ratio for NADH is 3).
Curve F is the same as E except that the P/O ratio for NADH is 2.

Table 4 presents predictions derived from the use of Figs. 2 and 3 for an obligate methanol-utilizer growing by way of the RuMP carbon assimilation pathway. It can be seen that the greatest potential increase in yield would be achieved by changing the methanol dehydrogenase to an NAD-linked dehydrogenase. It should be borne in mind that the very high yield that is predicted in this case might not be obtainable in practice on thermodynamic grounds. Such a modified organism might only be able to exist if it also 'adapted itself' by lowering the P/O ratio for NADH oxidation from 3 to 2. Even if this did occur then, as shown in Table 4, the predicted yield would still be higher than in the organism having the normal methanol dehydrogenase. In summery, bacteria growing on methanol by way of the RuMP pathway are some of the most efficient

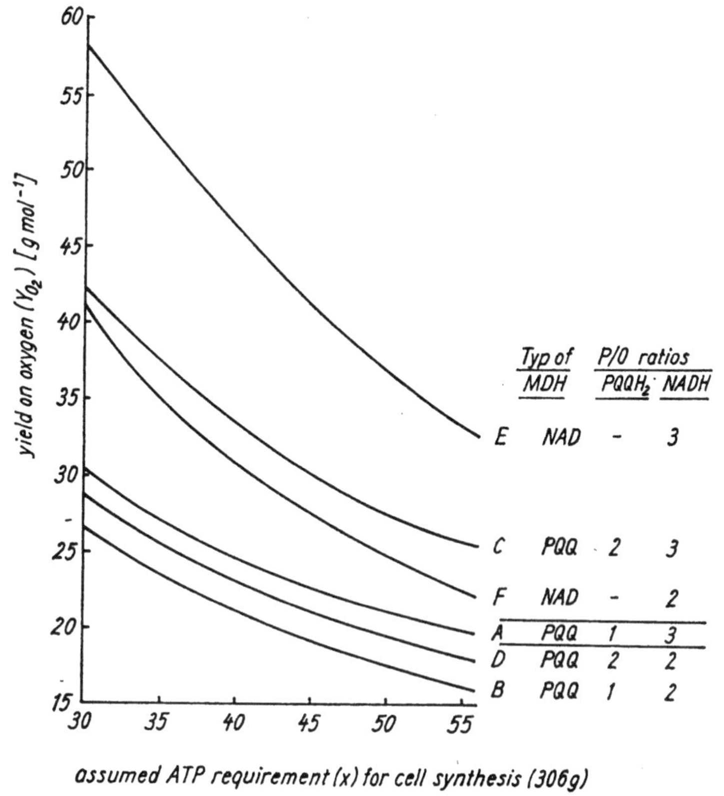

Fig. 3. The variation of Y_{O_2} with ATP requirement for growth

organisms known (in terms of carbon conversion efficiency), and the 'altered' organisms discussed here would be even more efficient. The free energy change during such efficient growth would tend to be much smaller than normal, and it must be concluded that rapid growth of such altered organisms might not be possible on thermodynamic grounds.

Table 4. Examples of changes in yields predicted from Figs. 2 and 3

These yields are for an obligate methanol-utilizer growing by way of the RuMP pathway of carbon assimilation. An ATP requirement of 37 ATP is assumed as the requirement for synthesis of 306 g of bacteria. The initial carbon conversion efficiency for such an organism would be 73% and the yield with respect to oxygen (Y_{O_2}) would be 26 g mol⁻¹.

Change in the organism proposed	Increases in yields	
	CCE (%)	Y_{O_2} (g mol⁻¹)
1. Increase in P/O ratio for methanol oxidation from the usual 1 to a value of 2		
if the P/O ratio for NADH is 3	73 to 83	26 to 36
if the P/O ratio for NADH is 2	67 to 73	23 to 25
2. Methanol dehydrogenase changes to NAD-linked enzyme		
if the P/O ratio for NADH is 3	73 to 95	26 to 49
if the P/O ratio for NADH is 2	67 to 82	23 to 33

References

[1] DUINE, J. A., FRANK, J.: In „Microbial Growth on C_1 Compounds“ (H. DALTON, ed.), pp. 31–41. Heyden, London, (1981).
[2] ANTHONY, C.: „The Biochemistry of Methylotrophs“, Academic Press, London, (1982).
[3] ANTHONY, C.: In „Microbial Growth on C_1 Compounds“ (H. DALTON, ed.), pp. 220–230. Heyden, London, (1981).
[4] O'KEEFFE, D. T., ANTHONY, C.: Biochem. J. **192** (1980) 411.
[5] CROSS, A. R., ANTHONY, C.: Biochem. J. **192** (1980) 421.
[6] OHTA, S., TOBARI, J.: J. Biochem. (Tokyo) **90** (1981) 215.
[7] BEARDMORE-GRAY, M., O'KEEFFE, D. T., ANTHONY, C.: Biochem. J. **207** (1982) 161.
[8] O'KEEFFE, D. T., ANTHONY, C.: Biochem. J. **190** (1980) 481.
[9] BEARDMORE-GRAY, M., O'KEEFFE, D. T., ANTHONY, C.: J. Gen. Microbiol., **129** (1983) 923.
[10] O'KEEFFE, D. T., ANTHONY, C.: Biochem. J. **170** (1978) 561.
[11] DAWSON, M. J., JONES, C. W.: Biochem. J. **194** (1981) 915.
[12] DAWSON, M. J., JONES, C. W.: Eur. J. Biochem. **118** (1981) 113.
[13] VAN VERSEVELD, H. W., KRAB, K., STOUTHAMER, A. H.: Biochim. Biophys. Acta **635** (1981) 525.
[14] ALEFOUNDER, P. R., FERGUSON, S. J.: Biochem. Biophys. Res. Comm. **98** (1981) 778.
[15] NETRUSOV, A. I., ANTHONY, C.: Biochem. J. **178** (1979) 353.
[16] ANTHONY, C.: J. Gen. Microbiol. **104** (1978) 91.
[17] ANTHONY, C.: In „Hydrocarbons in Biotechnology“ (D. E. F. HARRISON, I. J. HIGGINS & R. WATKINSON eds.), pp. 35–57. Heyden, London, (1980).

Acta Biotechnol. 3 (1983) 3, 269–277

Metabolic Features of Methane- and Methanol-Utilizing Bacteria

Y. A. Trotsenko

Institute of Biochemistry and Physiology of Microorganisms, USSR Academy of Sciences, Pushchino, Moscow region, 142292, USSR

Symposium on Biotechnology 1982, Leipzig, G.D.R.

Summary

Investigations of a wide range of methane- and methanol-utilizers showed a striking versatility of their metabolism dependent on the genotype and growth conditions. A correlation between pathways of carbon and nitrogen metabolism was found. It was most stringent in obligate methane-utilizers: the hexulosephosphate pathway bacteria assimilated NH_3 by the reductive amination of α-ketoglutarate or pyruvate whereas the serine pathway bacteria used the glutamate cycle (glutamine synthetase + glutamate-oxoglutarate aminotransferase).
Multiple enzymic lesions were found in central metabolism of obligate methylotrophs, i.e. the absence of the enzymes of glycolytic and pentosephosphate pathways, gluconeogenesis, citric acid cycle and glyoxylate shunt. These metabolic blocks were not so profound and could be compensated in restricted and facultative methylotrophs during heterotrophic growth.
The average levels of exogenous CO_2 fixation in methylotrophic bacteria with the hexulosephosphate, serine and ribulosebisphosphate pathways were found to be 10, 30 and 80% of their total cell carbon, respectively.
These results served as a basis for biotechnological applications of metabolic potential of methylotrophs (production of biomass, polysaccharides and enzymes as well as for microbiological treatment of industrial waters containing toxic C_1- and C_n-compounds).

Zusammenfassung

Untersuchungen an einer großen Anzahl von Methan und Methanol verwertenden Mikroorganismen zeigten, in welcher Vielfalt der Metabolismus vom Genotyp und von den Wachstumsbedingungen abhängig sein kann. Zwischen den Pathways des Kohlenstoff- und Stickstoffmetabolismus wurden Korrelationen gefunden. Dabei gilt für obligate Methan-Verwerter:
Hexulosephosphat-Weg-Bakterien assimilieren Ammoniak durch reduktive Aminierung von α-Ketoglutarat oder Pyruvat. Serin-Weg-Bakterien nutzen den Glutamat-Cyclus (Glutaminsynthetase und Glutamat-Oxoglutarat-Aminotransferase). Bei obligat methylotrophen Mikroorganismen wurde häufig das Fehlen von Enzymen zentraler Pathways wie der Glykolyse, des Pentosephosphat-Wegs, der Gluconeogenese, des Citrat-Cyclus und des Glyoxalat-Cyclus beobachtet. Diese metabolischen Blocks waren jedoch nicht so vollständig; sie konnten in restriktiven und fakultativen Methylotrophen unter heterotrophen Wachstumsbedingungen ausgeglichen werden. Die exogene CO_2-Fixierung trägt in methylotrophen Bakterien mit dem Hexulosephosphat-Weg zu 10%, mit dem Serien-Weg zu 30% und mit dem Ribulosediphosphat-Weg zu 80% zum Kohlenstoffgehalt der Zellen bei (Durchschnittswerte). Die genannten Ergebnisse können Grundlagen für die biotechnologische Anwendung der metabolischen Fähigkeiten von Methylotrophen z. B. bei der Biomasseproduktion, der Gewinnung von Polysacchariden und Enzymen bzw. der mikrobiologischen Behandlung von Industrieabwässern mit toxischen C_1- und C_n-Verbindungen sein.

1. Introduction

At present, methylotrophic microorganisms able to grow on methane and methanol are the object of extensive investigations. The prospects for biotechnological applications of methylotrophs are mainly based on the availability of pure and cheap methane and methanol as well as on minimal nutritional requirements of C_1-utilizers in comparison with heterotrophs. However, unfastidiousness of C_1-utilizers should not be regarded as a simplicity of their metabolism. Biochemical data convince that methylotrophs possess unique and versatile pathways for oxidation and assimilation of C_1-compounds.
A rapid progress in the problem of methylotrophy has been comprehensively reflected in literature, particularly in proceedings of four specialized symposia held in Tokyo (1974), Göttingen (1975), Pushchino (1977) and Sheffield (1980), as well as in recent book by Anthony "Biochemistry of methylotrophs" (1982). In some countries pilot plants for SCP production on methanol have been constructed. A new more promising direction is the production of various useful chemicals and biochemicals using immobilized cells and enzymes of methylotrophs. With the advent of recombinant DNA techniques the possibilities for construction of most efficient methylotrophic producers are indeed unlimited. Evidently, for effective application of methylotrophic potential a detailed knowledge of biology and metabolism of these fascinating microorganisms is needed. This paper describes some results of our studies on the metabolism of methane- and methanol-utilizing bacteria.

2. Pathways of primary C_1-assimilation

Aerobic methane- and methanol-utilizing bacteria can be conventionally divided into 3 main groups according to their pathways of primary C_1-assimilation. 1. Hexulosephosphate pathway bacteria ($3HCHO \rightarrow$ Triose). The representatives of the following genera belong to this group: *Pseudomonas, Methylomonas, Methylobacter, Methylococcus, Arthrobacter, Methylobacillus, Methylophilus, Acetobacter* [2, 3, 5, 6, 10—13, 16—20]. Although four variants of the hexulosephosphate cycle are possible, in fact only three of them are observed in bacteria tested. They include transaldolase or sedoheptulose 1,7-bisphosphatase rearrangements as well as aldolase cleavages via fructose-1,6-bisphosphate or 2-keto-3-deoxy-6-phosphogluconate. Usually, the hexulosephosphate pathway bacteria are widespread in nature and preferable, than other methylotrophs, for SCP production, in terms of their yields on methane and methanol.
2. Serine pathway bacteria ($2HCHO + CO_2 \rightarrow$ Triose). They belong to the following genera: *Methylosinus, Methylocystis, Pseudomonas, Hyphomicrobium*, and *Methylobacterium* [1, 3, 6, 10—13, 16—19]. These bacteria are also ubiquitous but have lower growth yields since the serine cycle is energetically less economic than the hexulosephosphate cycle. They can be additionally divided into two subgroups, having or lacking isocitrate lyase (icl^+ or icl^-).
3. Ribulosebisphosphate pathway bacteria ($3CO_2 \rightarrow$ Triose). In contrast to methylotrophs using the hexulosephosphate and serine pathways the bacteria of this group assimilate methanol after oxidation to CO_2. To label the presence of methylotrophic elements in their basic autotrophic metabolism a new term "pseudomethylotrophs" has been recently suggested by Zatman. Before 1977 only two species (*Paracoccus denitrificans* and *Thiobacillus novellus*) able to grow autotrophically on methanol and under $H_2/CO_2/O_2$ atmosphere were known [4, 8, 12]. Since 1977 we found such ability in several bacteria belonging to very different genera: *Pseudomonas, Microcyclus, Achromobacter, Mycobacterium, Blastobacter, Renobacter, Angulomicrobium* [4, 8, 17, 18]. Thus, despite the ribulosebisphosphate (RBP) cycle is most energetically unfavourable this type of methanol metabolism may not be rare.

It should be noted that the above division of methane and methanol-utilizers into 3 groups is rather conventional because in some bacteria several pathways of primary C_1-assimilation can operate simultaneously. For instance, in obligate methanotroph *Methylococcus capsulatus* assimilating formaldehyde via the hexulosephosphate cycle a minor contribution of the RBP and serine pathways has been reported [11, 12]. This fact points to the metabolic flexibility and evolutionary close-relatedness of methylotrophic and autotrophic bacteria.

3. Pathways of intermediary metabolism and problem of obligate methylotrophy

Many methane- and methanol-utilizing bacteria are obligate methylotrophs. Possible reasons for their obligate dependence on C_1-substrates, as in case of strict autotrophs, are the metabolic blocks in energetic and biosynthetic pathways [10]. For instance, Type I obligate methanotrophs, with the hexulosephosphate pathway of formaldehyde assimilation, possess the incomplete Krebs cycle lacking α-ketoglutarate dehydrogenase. As a result they are unable to oxidize and grow on polycarbon compounds. In contrast, Type II obligate methanotrophs, with the serine pathway, have a complete set of the Krebs cycle enzymes but lack hexokinase and 6-phosphogluconate dehydrogenase

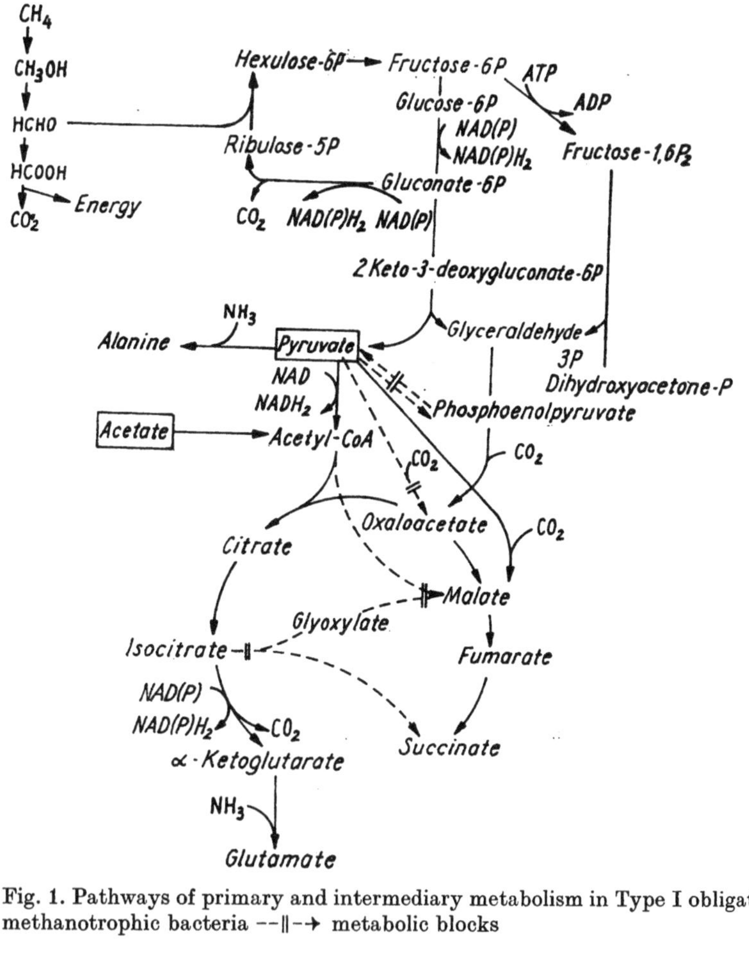

Fig. 1. Pathways of primary and intermediary metabolism in Type I obligate methanotrophic bacteria --||--> metabolic blocks

However, the absence of α-ketoglutarate dehydrogenase in Type I methanotrophs appears to be an insufficient reason for explaining their failure to grow on multicarbon compounds, since this enzyme may be completely repressed in typical heterotrophic bacteria, e.g., *Escherichia coli* growing anaerobically on glucose. The cause for obligateness of Type II methanotrophs is also not clear, since the lack of hexokinase and 6-phosphogluconate dehydrogenase only impairs growth of these bacteria on sugars. In order to explore this problem of obligateness further, we set out to analyze possible metabolic routes of pyruvate and acetate in both types of obligate methanotrophs [14] as well as in obligate and facultative methanol-utilizers [2, 5]. Measurements of the metabolic activity of cell suspensions an dextracts have shown that both substrates are utilized by the bacteria tested. However, the possibilities for their (i.e. pyruvate and acetate), channeling into major energetic and biosynthetic pathways are very limited. This conclusion is consistent with the original findings of other workers who have shown that, although dependent upon a C_1 source, acetate can increase markedly the growth yield but is incorporated only into lipid and amino acid fractions in obligate methanotrophs [10].

In Type I methanotrophs, having the hexulosephosphate pathway, pyruvate and acetate can enter both the oxidative branch of the Krebs cycle through acetyl-CoA and, to a lesser extent, its reductive branch after carboxylation by "malic" enzyme (Fig. 1). However, owing to the absence of α-ketoglutarate dehydrogenase, pyruvate and acetate cannot be completely oxidized and so serve as an energy source. Moreover, these bacteria

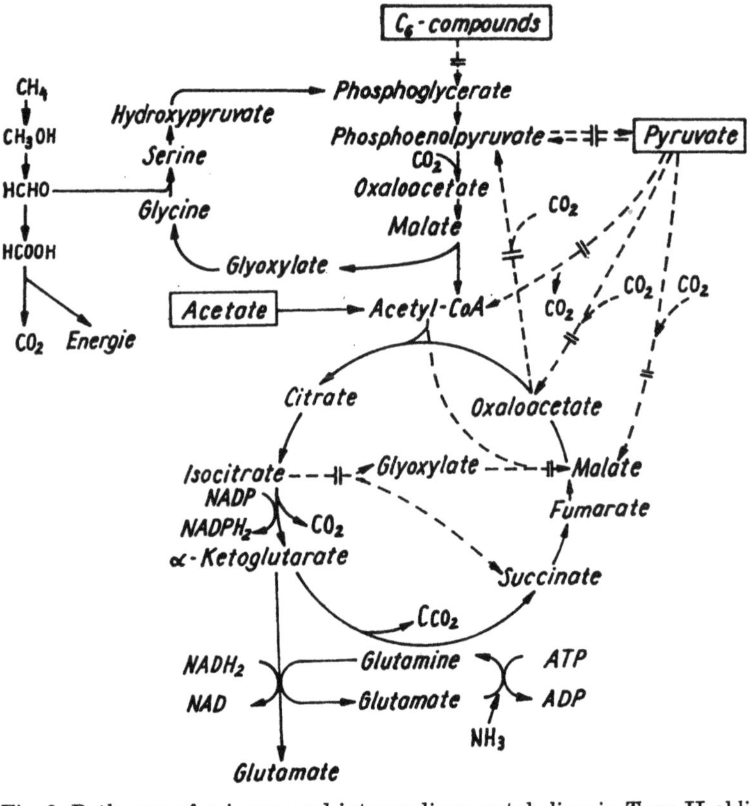

Fig. 2. Pathways of primary and intermediary metabolism in Type II obligate methanotrophic bacteria --||-→ metabolic blocks

do not possess the activities of glyoxylate shunt specific enzymes — isocitrate lyase and malate synthase. These metabolic defects appear to explain the C_1-compound obligateness of these organisms.

In Type II methanotrophs, with the serine cycle, pyruvate metabolism via acetyl-CoA is of minor importance due to the extremely low activity of E_1-component of the pyruvate dehydrogenase complex (Fig. 2). The enzymes necessary for anaplerotic replenishment of the Krebs cycle intermediates, viz., isocitrate lyase and malate synthase, are absent. In these bacteria PEP can not rise from pyruvate because of the lack of PEP-synthetase and pyruvate phosphate dikinase. The other route involving pyruvate carboxylase and PEP-carboxykinase is also inoperative. Thus, in Type II obligate methanotrophs all known gluconeogenic pathways from pyruvate and acetate are blocked.

It should be also noted that absence of pyruvate kinase in both types of obligate methanotrophs has a general physiological significance since it makes impossible for them to convert PEP to pyruvate and to generate energy from carbon compounds which are metabolized via glycolytic sequence.

The profiles of some essential enzymes of intermediary carbon metabolism in obligate (*Methylophilus methanolovorus*) and facultative (*Pseudomonas oleovorans*) methylotrophs are given in Table 1. It follows that in both organisms the activities of gluconeogenic enzymes (fructosebisphosphate aldolase, fructosebisphosphatase, PEP-carboxy-

Table 1. Levels of the Enzymes of Intermediatory Carbon Metabolism in Obligate and Facultative Methanol-Utilizing Bacteria with Hexulosephosphate Cycle of Formaldehyde Assimilation

Enzyme		Enzyme activities (n mol/min/mg protein)		
		Methylophilus methanolovorus	*Pseudomonas oleovorans*	
		Methanol	*Methanol*	*Acetate*
Hexokinase		0	4	6
Glucose-6-phosphate dehydrogenase				
	NAD	4785	306	23
	NADP	5742	1076	81
Gluconate-6-phosphate-dehydrogenase	NAD	957	64	20
	NADP	1436	260	30
KDPG-aldolase		157	130	8
Fructose-1,6-bisphosphate aldolase		0	0	27
Fructose-1,6-bisphosphatase		0	0	78
Sedoheptulose-1,7-bisphosphatase		0	0	0
PEP synthetase		0	2	4
Pyruvate kinase		161	163	201
Pyruvate dehydrogenase		16	20	191
Pyruvate carboxylase		48	0.7	0.2
PEP-carboxylase		26	0.1	0.9
PEP-carboxykinase		0	0.1	1.1
Pyruvate phosphate dikinase		0	0	0
α-Ketoglutarate dehydrogenase		0	0	32
Isocitrate lyase		19	18	29
Malate synthase		20	14	13

KDPG — 2-keto-3-deoxy-6-phosphogluconate; PEP — phosphoenolpyruvate.

kinase) as well as α-ketoglutarate dehydrogenase are very low during growth on methanol. In contrast to *M. methanolovorus*, all these enzymes are induced or derepressed in *Ps. oleovorans* when grown on acetate. Hence, our results indicate that the irreversible loss of several enzymes necessary for heterotrophic growth, rather than a single enzymic lesion, along with the operation of highly specialized C_1-metabolism may be a biochemical basis of obligate methylotrophy [2, 5, 14].

4. Assimilation of exogenous CO_2

Although carbon dioxide plays a very important role in the metabolism of methylotrophic bacteria the estimation of CO_2 contribution to their cell material synthesis is still controversial. It has been reported that *Pseudomonas* AM 1, having the serine pathway, derives 50% of the cell carbon from CO_2 during growth on methanol [1]. Based on different activities of pyruvate and phosphoenolpyruvate carboxylases we proposed that serine pathway methylotrophs assimilate much more CO_2 than hexulosephosphate pathway bacteria [6, 16]. However, recently Malashenko et al [10, 11] have reported that all methanotrophic bacteria, independently on their pathways of primary C_1-assimilation, fix equal quantity of exogenous CO_2, viz., 30%. Exploring this point again we determined by radioisotopic method the contribution of CO_2 to cell material synthesis in growing cultures of different methylotrophic bacteria.

Table 2. Levels of Exogenous CO_2 Fixation by Methane- and Methanol-Utilizing Bacteria During Methylotrophic Growth (% of total cell carbon)

Organisms	Pathways of primary C_1-assimilation		
	Hexulosephosphate cycle	Serine cycle	Ribulose-bisphosphate cycle
Methanotrophs	5—10	30	8 (*Mc. capsulatus*)
	PEPC, ME	PEPC	RBPC, PEPC
Methylotrophs	12—15 PC, PEPC, ME, (PEPK)	40—50 PEPC	80 RBPC, PC, PEPC, PEPK

PEPC — Phosphoenolpyruvate carboxylase, PC — pyruvate carboxylase,
PEPK — phosphoenolpyruvate carboxykinase, ME — "malic" enzyme,
RBPC — ribulosebisphosphate carboxylase

As seen from Table 2 various groups of methane- and methanol-utilizing bacteria differ clearly by their levels of CO_2 assimilation. Hexulosephosphate pathway methanotrophs derive 5—10% of total cell carbon from CO_2 whereas those with the serine pathway —30%. PEP-carboxylase (EC 4.1.1.31) is mainly responsible for CO_2 fixation in most methanotrophs, the activity of this enzyme being much higher in serine pathway bacteria. Only in *Methylococcus capsulatus* the main CO_2-fixing enzyme appeared to be RBP-carboxylase. No activity of pyruvate carboxylase was found in all the methanotrophs studied.

The levels of CO_2 fixation were somewhat higher in methanol-utilizing bacteria. So, hexulosephosphate pathway methanol-utilizers derive 12—15% of the cell carbon from CO_2, whereas those with the serine pathway — 50% and RBP pathway bacteria — 80%. It should be emphasized that methanol-utilizing bacteria also differed from metha-

notrophs by a higher variety of carboxylases, particularly, by the presence of pyruvate carboxylase which was the most active in hexulose-phosphate pathway methanol-utilizers.

Our results are in good accordance with the equations of primary C_1-assimilation [12] and the data of other workers who showed a stronger stimulatory effect of exogenous CO_2 on cell yield of serine pathway methanotrophs as compared with those using hexulose-phosphate pathway [19]. In principle we agree with the statement of MALASHENKO et. al. [11] that CO_2-fixing potential in microorganisms have to be determined by the reduction level of growth substrates, i.e. by the quantity of electrons available for acception by CO_2. However, our data strongly suggest that the actual scale of CO_2 fixation is mainly dependent on growth conditions and the type of metabolism of methane- and methanol-utilizers.

Hexulosephosphate pathway bacteria possess the lowest level of carbon dioxide fixation (5—15%) since CO_2 is not involved in the primary synthesis of trioses in these organisms. They assimilate CO_2 exclusively for anaplerotic replenishment of the Krebs cycle intermediates withdrawn for biosynthesis because of the absence of α-ketoglutarate dehydrogenase. Serine pathway bacteria assimilate carbon dioxide much more intensively (30—50%) since derive trioses from two molecules of formaldehyde and one molecule of CO_2. So, at least one third of their cell carbon should come from CO_2. RBP pathway bacteria have the highest level of CO_2 fixation (80%). Theoretically all their cell carbon should derive from CO_2, for instance during autotrophic growth under $H_2/CO_2/O_2$ atmosphere. It is reasonable to assume that these bacteria during growth on methanol assimilate this substrate not only at the level of CO_2 but also in a more reduced form, i.e. as tetrahydrofolate derivatives of formaldehyde or formate via hydroxymethylation of glycine to serine. This is confirmed by the presence of highly active hydroxypyruvate reductase, which is the specific enzyme of the serine pathway [4, 8].

It is interesting that the levels of CO_2 fixation are extremely low in the bacteria tested during heterotrophic growth (2—4%). This can be explained by derepression of α-ketoglutarate dehydrogenase and amphibolic function of the Krebs cycle.

Thus, we believe that CO_2-fixing potential of methane- and methanol-utilizing bacteria depends mainly on their genotype (type of metabolism) and phenotype (growth conditions) than on reduced level of growth substrates. This conclusion is supported by higher percentage of CO_2 fixation and the activities of appropriate carboxylases in methanol — than methane — grown bacteria and by the minimal percentage of CO_2 fixation in heterotrophically grown cells. It follows that exogenous CO_2 can be effectively used as an additional source of carbon during large scale cultivation of methane- and methanol-utilizers.

5. Pathways of ammonia assimilation

We have proposed that the previously established correlation between the types of intracytoplasmic membranes (ICM) and pathways of primary and intermediary metabolism in obligate methanotrophs [10, 16] should also be extended to include ammonia assimilatory routes. With this aim in view many strains of methane- and methanol-utilizing bacteria were examined for the occurence of enzymes responsible for the primary utilization of ammonia with a variety of nitrogen sources. Our results showed that the obligate methanotrophs tested actually differ in the pathways of ammonia assimilation [13].

As seen from Table 3 methanotrophs, having Type I of ICM and the hexulosephosphate pathway, assimilate NH_4 by the reductive amination of α-ketoglutarate or pyruvate and via glutamine synthetase. Alternatively, methanotrophs, having Type II of ICM

and the serine pathway, assimilate ammonia exclusively by the enzyme system of the glutamate cycle, namely glutamine synthetase and glutamate synthase. Only *Mc. capsulatus*, having Type I of ICM and the hexulosephosphate pathway with minor contribution of the RBP and serine pathways, occupies an intermediate position since this organnism uses both the glutamate cycle and alanine dehydrogenase for ammonia assimilation.

Table 3. Distribution of Primary Ammonia-Assimilating Enzymes among Methane- and Methanol-Utilizing Bacteria

Organisms	Pathways of primary C_1-assimilation		
	Hexulosephosphate cycle	Serine cycle	Ribulosebisphosphate cycle
Methanotrophs	Dehydrogenases of glutamate and alanine	GS/GOGAT	GS/GOGAT Alanine dehydrogenase (*Mc. capsulatus*)
Methylotrophs	Dehydrogenases of glutamate and alanine GS/GOGAT	GS/GOGAT Dehydrogenases of alanine and glycine	GS/GOGAT Alanine dehydrogenase

GS/GOGAT — glutamine synthetase + glutamate—oxoglutarate aminotransferase.

Such correlation between carbon and nitrogen assimilatory pathways was not so distinct in methanol-utilizing bacteria [3]. Some hexulosephosphate pathway methanol-utilizers assimilate NH_4 by reductive amination of α-ketoglutarate and pyruvate others-via the glutamate cycle. Serine pathway methanol-utilizers use the glutamate cycle for NH_4 assimilation as well as reductive amination of pyruvate or glyoxylate. All RBP pathway bacteria tested can assimilate ammonia via the glutamate cycle or by reductive amination of pyruvate. The reason for simultaneous operation of several ammonia assimilatory routes in some methanol-utilizers is not clear. Perhaps it reflects their higher metabolic flexibility than that of methanotrophs.

In this connection let me just remind you of the elegant experiments of ICI group in England on introduction of the glutamate dehydrogenase gene of E. coli into obligate methanol-utilizer *Methylophilus methylotrophus* [20]. Assimilation of ammonia via glutamate dehydrogenase is more energy-efficient than via the glutamate cycle, thus the recombinant organism converts more growth substrate, methanol, into cellular carbon. This represents a new and attractive area for the application of recombinant DNA techniques for industrial and agricultural use of metabolic potential of C_1-utilizers.

6. Concluding remarks

Metabolic features discovered in methane and methanol-utilizing bacteria served us as a basis for their biotechnological applications in production of biomass, polysaccharides and enzymes as well as for microbiological treatment of industrial waters containing toxic C_1-and C_n-compounds [9, 15, 18]. Despite these achievements much more information, particularly with respect to the metabolic and genetic regulation of the pathways involved in methane and methanol utilization is required.

References

[1] Large, P. J., Peel, D., Quayle, J. R.: Biochem. J. **81** (1961) 470.

[2] Loginova, N. V., Govorukhina, N. I., Trotsenko, Y. A.: Mikrobiologiya **50** (1981) 305.

[3] Loginova, N. V., Govorukhina, N. I., Trotsenko, Y. A.: Mikrobiologiya **51** (1982) 38.

[4] Loginova, N. V., Namsaraev, B. B., Trotsenko, Y. A.: Mikrobiologiya **47** (1978) 168.

[5] Loginova, N. V., Trotsenko, Y. A.: Mikrobiologya **46** (1977) 210.

[6] Loginova, N. V., Trotsenko, Y. A.: Mikrobiologya **48** (1979) 202.

[7] Loginova, N. V., Trotsenko, Y. A.: Mikrobiologya **48** (1979) 785.

[8] Loginova, N. V., Trotsenko, Y. A.: FEMS Microbiology Letters **5** (1979) 239.

[9] Loginova, N. V., Trotsenko, Y. A.: Appl. Biochem. Microbiol. **16** (1980) 331.

[10] Malashenko, Y. R., Romanovskaya, V. A., Trotsenko, Y. A.: Methane-oxidizing microorganisms (Ed. by G. A. Zavarzin) pp. 1—109. Nauka, Moscow, 1978.

[11] Romanovskaya, V. A., Luydvichenko, E. S., Kryshtab, T. P., Zhukov, V. G., Sokolov, I. G., Malashenko, Y. R.: Mikrobiologiya **49** (1980) 687.

[12] Quayle, J. R., Ferenci, T.: Microbiol. Rev. **42** (1978) 251.

[13] Shishkina, V. N., Trotsenko, Y. A.: FEMS Microbiology Letters **5** (1979) 187.

[14] Shishkina, V. N., Trotsenko, Y. A.: FEMS Microbiology Letters **13** (1982) 237.

[15] Sokolov, A. P., Luchin, S. V., Trotsenko, Y. A.: Biokhimiya **45** (1980) 137.

[16] Trotsenko, Y. A.: Microbial Production and Utilization of Gases (Ed. by H. G. Schlegel, G. Gottschalk and N. Pfennig) pp. 329—336. Goltze, Göttingen, 1976.

[17] Trotsenko, Y. A., Loginova, N. V.: Uspekhi mikrobiologii **14** (1979) 28.

[18] Trotsenko, Y. A., Loginova, N. V.: Microbiological methods of wastes treatment. Abstracts of the Proceedings of All-Union Conference. p. 113. Pushchino, 1980.

[19] Wendlandt, K.-D., Brühl, E.: Abstracts of 2nd Symp. Soc. Countries on Biotechnology, p. 113, Leipzig, 1980.

[20] Windass, J. D., Worsey, M. J., Pioli, E. M., Pioli, D., Barth, P. T. Atherton, K. T., Dart, E. C., Byrom, D., Powell, K., Senior, P. J.: Nature **287** (1980) 396.

Acta Biotechnol. 3 (1983) 3, 278

BOOK REVIEW

New Trends in Research and Utilization of Solar Energy through Biological Systems.

Edited by H. MYSLIN und R. BACHOFEN. Birkhäuser Verlag Basel—Boston—Stuttgart 1982, 156 S.

Das vorliegende Buch bildet den Nachdruck von im Jahre 1982 in Experientia veröffentlichten Übersichtsartikeln.

Nach einem einführenden Überblick über das Gesamtproblem von D. O. HALL, der u. a. die Vorstellungen des Autors zur Entwicklung von „Biomasse für Brennstoff" — Programmen bis über das Jahr 2000 hinaus enthält, werden in den einzelnen Kapiteln zu den Themen „Höhere Pflanzen als Energiekonverter", „Wasserpflanzen und Algen als Energiekonverter", „Biologische Photoproduktion von Ammoniak und Wasserstoff", „Konversion von Biomasse in Brennstoffe und Grundchemikalien", „Bildung von Methan aus Biomasse — Ökologie, Biochemie und Applikation" und „Zukünftige Systeme" von namhaften Autoren Stand und Tendenzen auf dem Gebiet der photobiologischen Solarenergienutzung behandelt. Das Spektrum reicht dabei vom Anbau Kohlenwasserstoffe enthaltender Pflanzen bis zur Photoproduktion von Wasserstoff mit Hilfe von lebenden Zellen und von am Photosyntheseapparat orientierten artifiziellen Systemen.

Das Buch schließt mit der nur selten angebotenen Darstellung aller photobiologischer Möglichkeiten zur Solarenergiekonvertierung einschließlich der Sekundärproduktion von Nahrungs- und Futtermitteln, von Energieträgern und Grundchemikalien eine Lücke zwischen der in der Vergangenheit häufigeren Behandlung der Algenkultivierung und -nutzung und der umfassenden Darstellung der die Welt bedrückenden Nahrungs-, Energie- und Rohstoffsorgen, in der die Potenzen der Photobiologie lediglich am Rande erwähnt werden.

Die auf engem Raum angebotene Fülle, nicht zuletzt von ökonomischen Daten, könnte sich für eine Abschätzung der Realisierungsaussichten der einzelnen Verfahren als außerordentlich hilfreich erweisen, auch wenn sie sich leider auf Grund unterschiedlicher Bezugsgrößen teilweise nur schwer miteinander vergleichen lassen, was infolge der heterogenen Zusammensetzung der Autoren sicher unvermeidlich ist.

P. ROTH

Acta Biotechnol. 3 (1983) 3, 279—288

Regulation der Fettsäurebildung bei der mikrobiellen Alkanoxidation

H. J. Rehm, L. Hortmann und I. Reiff

Institut für Mikrobiologie, Universität Münster
Tibusstraße 7—15, D-4400 Münster

Biotechnologie-Symposium 1982, Leipzig, DDR

Summary

At the microbial oxidation of long-chain alkanes such monocarbonic acids at a monoterminal oxidation are formed which correspond to the chain length of the alkanes. At diterminal oxidation the corresponding dioic acids are produced.
Under the influence of cerulenin a direct incorporation of fatty acids into cell lipids increases. Undecanoic acid cannot be metabolized by *Mortierella isabellina.* It causes an inhibition of alkane oxidation. A main effect of undecanoic acid is an inhibition of the elongation of other fatty acids. The de novo fatty acid biosynthesis has not been inhibited by undecanoic acid.

Zusammenfassung

Bei der mikrobiellen Oxidation längerkettiger Alkane entstehen bei einer terminalen Oxidation die der Kettenlänge entsprechenden Monocarbonsäuren, bei einer diterminalen Oxidation werden entsprechende Dicarbonsäuren gebildet.
Unter dem Einfluß von Cerulenin erhöht sich der direkte Einbau der Fettsäuren in die Zellipide. Undekansäure (Hendecylsäure) wird von *Mortierella isabellina* nicht als Fettsäure aufgenommen, sondern ruft eine Hemmung der Alkanoxidation hervor. Eine wesentliche Wirkung der Undekansäure liegt in einer Hemmung der Elongation anderer gebildeter Fettsäuren. Die de novo-Fettsäurebiosynthese wird durch Undekansäure nicht gehemmt.

Einleitung

Über die Regulation der mikrobiellen Alkanoxidation gibt es eine Reihe von Ergebnissen: Rehm und Reiff [10, 11]; Kleber und Sattler [6]. Dagegen ist die Regulation der Bildung von aus der Alkanoxidation entstandenen Folgeprodukten, besonders der als erste wichtige Substanzen entstandenen Fettsäuren, weniger untersucht worden, obwohl es eine ganze Anzahl von Arbeiten gibt, in denen die aus Alkanen gebildeten extrazellulären und intrazellulären Fettsäuren identifiziert worden sind.
Im folgenden sollen einige wichtige Mechanismen der Regulation der Fettsäurebildung dargestellt werden und zwar

1. die Regulation in Abhängigkeit von Oxidationsmechanismen
2. Kinetiken der Fettsäurebildung monoterminal oxidierender Mikroorganismen
3. Wirkung von Cerulenin auf die Fettsäurebildung von *Mortierella isabellina*
4. besondere Regulation durch eine C_{11}-Säure.

Material und Methoden

1. Regulation der Fettsäurebildung in Abhängigkeit von den Oxidationsmechanismen

Es ist gut bekannt, daß Bakterien, Hefen und auch Schimmelpilze die aus der Oxidation von Alkanen — besonders von längerkettigen Alkanen — gebildeten organischen Säuren durch β- oder auch gelegentlich durch α-Oxidation abbauen. Dieser Abbau scheint allgemein der Regulation der genannten Oxidationsmechanismen zu unterliegen.
Weiterhin ist bekannt, daß ein Teil der aus Alkanen gebildeten langkettigen Fettsäuren direkt in die Zellipide eingebaut wird.

1. Fettsäurebildung beim monoterminalen Oxidationsmechanismus

Bei diesem Oxidationsmechanismus werden — entsprechend der Länge der Alkane — nicht nur geradkettige, sondern auch ungeradkettige Fettsäuren in die Zellipide eingebaut. Parallel dazu kann entweder eine β-Oxidation stattfinden, so daß man die nun jeweils um 2 C-Atome verkürzten Fettsäuren in den Zellipiden findet. Bei einer Elongation andererseits kann man nun die 2 C-Atome oder ein Mehrfaches von 2 Atomen längeren Fettsäuren nachweisen. Bei Applikation von ungeradkettigen Alkanen oder auch ungeradkettigen primären Oxidationsprodukten, z. B. primären Alkoholen als Substrat, ist dieser Effekt besonders gut — ohne die Verwendung markierter Alkane — zu erkennen.

2. Fettsäurebildung beim diterminalen Oxidationsmechanismus

Da die aus einer Alkanoxidation entstandenen Disäuren nicht in die Zellipide eingebaut werden, kann man die Oxidationsprodukte im wesentlichen nur extrazellulär im Substrat erkennen. Auch hier ist eine β-Oxidation jeweils an einer Seite möglich. Nach der Aufklärung einer diterminalen Oxidation über das entsprechende Diol (Yi und Rehm [15]), ist auch eine gleichzeitige β-Oxidation an beiden Enden des Moleküls zu erwarten.
Sowohl bei der mono- als auch bei der diterminalen mikrobiellen Alkanoxidation wird die freie Alkansäure in Cytosol entweder mit HS-CoA durch die langkettige Acyl-CoA-Synthetase (System I) gekoppelt, dann elongiert und in das Lipid direkt eingebaut oder mit Hilfe der langkettigen Acyl-CoA-Synthetase (System II) bei Pilzen in den Mitochondrien oder den Peroxysomen mit HS-CoA gekoppelt und dann β-oxidiert.
Dabei kann schon nach einmaligem bis mehrfachem Durchlaufen der β-Oxidation ein Einbau in die Zellipide erfolgen.
Daneben erfolgt zweifellos immer eine vollständige β-Oxidation zu Acetyl-CoA. Bei Eukaryonten wird die C_2-Einheit nach Abspaltung des Acetyl-CoA mit Hilfe der kurzkettigen CoA-ACP-Transferase als Acetylrest in die de novo-Fettsäuresynthese eingeschleust. Evtl. ist hierbei ein Acetyl-carnitin-System beteiligt.
Anscheinend werden bei großem Angebot an Alkanen besonders viel Fettsäuren direkt aus der Alkanoxidation in die Zellipide eingebaut.

3. Fettsäurebildung bei subterminalen Oxidationsmechanismen

Bei den verschiedenen Typen der mikrobiellen subterminalen Alkanoxidation wird das Molekül in Bruchstücke verschiedener Längen geteilt, so daß Fettsäuren sehr unterschiedlicher Länge entstehen. Aussagen über den direkten Einbau dieser Fettsäuren oder den vollständigen Abbau zu C_2-Einheiten und eine de novo-Synthese der Fettsäuren

können nicht gemacht werden. Ebenso ist es nicht möglich, über die Regulation etwas auszusagen. Möglicherweise erfolgt vielfach eine de novo-Synthese, da die Fettsäuresequenzen denen der auf Glukose gewachsenen Mikroorganismen sehr ähnlich sind.

II. Kinetik der Bildung von Fettsäuren bei monoterminal oxidierenden Mikroorganismen

Bisher wurden kaum Änderungen der Fettsäurezusammensetzung in Zellipiden im Verlauf der Mikroorganismenentwicklung untersucht.

Werden die Zellen, nachdem sie auf Alkanen gewachsen sind, nicht mehr mit Alkanen versorgt, so vermindern sich die Gesamtfettsäuren relativ schnell. Auch die Zusammensetzung der einzelnen Fettsäuren ändert sich im Laufe der Zeit nicht unwesentlich (Abb. 1).

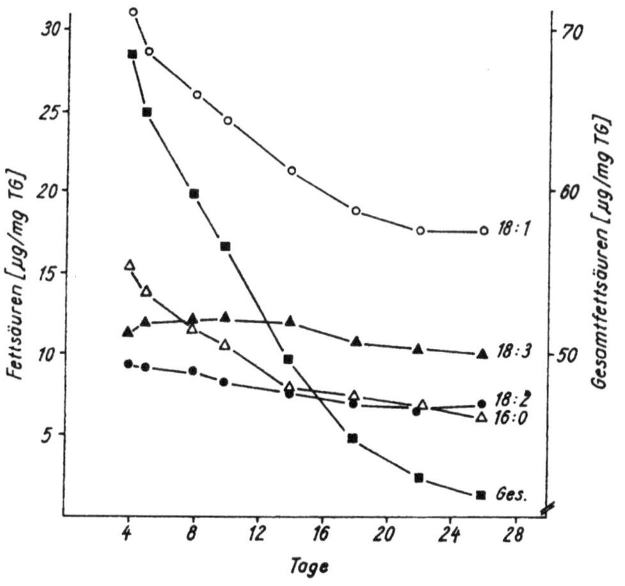

Abb. 1. Veränderung der Fettsäuren bei hungernden Zellen von *Mortierella isabellina* nach Wachstum auf Alkanen

Eindrucksvoll ist die Bildung von γ-Linolensäure auf Glukose und Tetradekan im Verlauf von 60 Tagen. Sie ist relativ zu den verschiedenen Fettsäuren sehr unterschiedlich, absolut aber sowohl auf Glukose als auch auf Tetradekan nach einiger Zeit auf eine erstaunlich gleichmäßige Menge pro Trockengewicht eingependelt (Abb. 2).

III. Wirkung von Cerulenin auf die Fettsäurebiosynthese

Cerulenin (2,3-Epoxy-4-oxo-6,10-dodecadienoylamid) hemmt die de novo-Fettsäuresynthese auf der Stufe der β-Ketoacylthioestersynthetase (condensing enzyme) (Nomura et al. [9]; Vance et al. [14]).

Es sollte bei dem vorwiegend monoterminal oxidierenden Pilz *Mortierella isabellina* untersucht werden, ob dieser in der Lage ist, bei Blockierung der Fettsäuresynthetase die notwendigen Fettsäuren direkt aus dem Substrat zu übernehmen. Der Hemmstoff

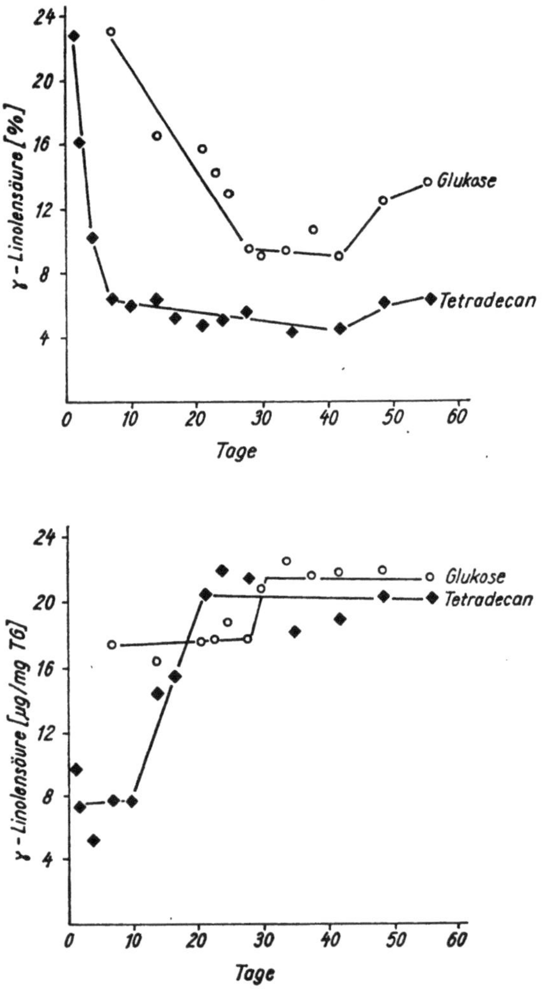

Abb. 2. Bildung von γ-Linolensäure durch *Mortierella isabellina*

beeinträchtigt nicht die Verbindung der Fettsäure mit HS-CoA und die Elongation (Tanaka et al. [12]). Eine ausführliche Beschreibung der Wirkung von Cerulenin auf *M. isabellina* vgl. Hortmann und Rehm [2].

a) *Wirkung von Cerulenin mit Glukose als Substrat*

Die Wirkung von Cerulenin auf *M. isabellina* bei Glukose als Substrat zeigt die Tab. 1. Sie zeigt, daß keine vollständige Hemmung durch Cerulenin zu erreichen ist, daß aber bereits ab 5 μg/l Nährlösung eine Hemmung der de novo-Fettsäuresynthese eintritt. Es ist durch die Hemmung der de novo-Synthese eine Minderung des Trockengewichts eingetreten. Das Fettsäuremuster zeigt nur die geradzahligen Fettsäuren (Abb. 3).

b) *Wirkung mit Alkan als Substrat*

Bei der Applikation von Cerulenin mit Alkanen als C-Quelle ergeben sich folgende Schwierigkeiten:

Die wirksame Ceruleninkonzentration ist anders als in wäßriger Lösung, denn durch die Alkane wird das lipophile Antibiotikum aus der wäßrigen Lösung extrahiert und direkt

Tabelle 1. Trockengewichte von *Mortierella isabellina* nach Wachstum auf 4% Glukose und Ceruleninzusatz

Konz. an Cerulenin µg/ml	Inkubationszeit Tage	Trockengewicht mg
0,0	18	207,2
5,0	18	113,5
10,0	18	64,2
20,0	24	48,4
30,0	26	41,3
40,0	28	39,7
50,0	30	35,0
75,0	33	33,1
100,0	36	29,4

mit dem Alkansubstrat in höheren Konzentrationen als aus der wäßrigen Phase aufgenommen. Daher war der Pilz auch schon gegen relativ niedrige Konzentrationen empfindlich. Die Inkubationszeiten waren sehr verlängert.
Nach Wachstum auf geradzahligen Alkanen wie Hexa- und Octadekan zeigte der Pilz ein ähnliches Fettsäuremuster wie auf Glukose. Besonders C_{16}- und C_{18}-Fettsäuren waren gebildet worden.

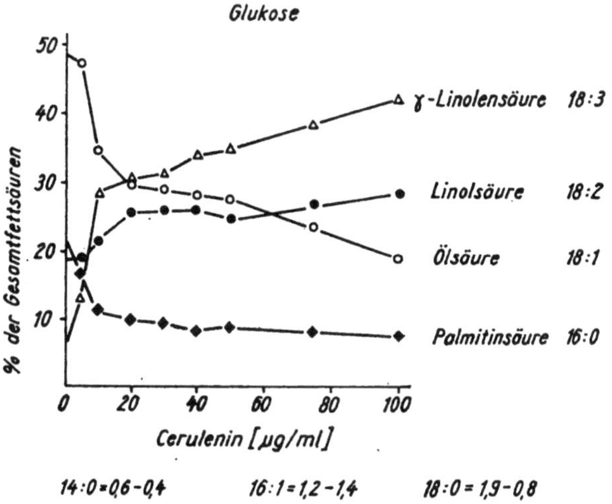

Abb. 3: Fettsäuremuster bei *Mortierella isabellina* nach Wachstum auf Glukose bei verschiedenen Cerulenin-Konzentrationen

Tridekan war wesentlich interessanter: Je höher die Ceruleninkonzentration war, desto mehr fand ein direkter Einbau der ungeradkettigen Fettsäuren in die Lipide statt. Die Tab. 2 zeigt besonders eine Anhäufung von C_{17}-, C_{19}-, aber auch von C_{15}- und C_{13}-Fettsäuren. Je stärker die de novo-Synthese gehemmt wird, desto stärker wird also auch der direkte Fettsäureeinbau begünstigt. Die β-Oxidation läuft anscheinend weniger ab.

c) *Wirkung mit Citrat und Pentadekan*

Die Hemmung durch Cerulenin wird in nicht geringem Maße durch das Substrat bestimmt. Um dem Pilz eine bessere Entwicklung als bei reinem Alkansubstrat zu ermög-

Tabelle 2. % der Fettsäuren bei *Mortierella isabellina* nach Wachstum auf Tridekan bei Ceruleninzusatz

	Konzentration an Cerulenin µg/ml nl:				
	0,0	10,0	20,0	35,0	50,0
Inkubationszeit Wochen:	5	5	6	8	10
Fettsäuren					
13:0	3,0	3,4	5,0	6,8	6,8
14:0	0,6	0,4	0,3	0,2	0,2
15:0	5,3	5,9	8,0	7,1	10,7
15:1	—	—	Sp	Sp	0,1
16:0	13,9	12,1	10,8	2,4	2,4
16:1	0,7	0,7	0,3	0,8	0,9
17:0	0,1	0,2	0,3	1,6	2,1
17:1	6,2	10,4	16,5	27,8	35,7
17:2	0,2	0,9	1,6	4,2	4,3
17:3	0,1	0,2	0,2	0,3	0,4
18:0	0,2	0,2	0,1	Sp	0,1
18:1	41,8	39,2	31,3	23,8	17,4
18:2	16,4	14,3	12,9	10,2	5,6
18:3	11,4	11,0	10,9	8,4	6,4
19:0	Sp	0,1	0,1	0,2	0,3
19:1	0,1	0,9	2,0	5,9	6,2
19:2	Sp	0,1	0,2	0,3	0,3

lichen, mußte eine weitere C-Quelle zugesetzt werden. Da Glukose als Inhibitor des Alkanabbaus nicht infrage kam, wurde Citrat, das vom Pilz verwertet wird, als zusätzliche C-Quelle zu Pentadekan verwendet. Der Konzentrationseffekt des Alkans bleibt natürlich bestehen, so daß ein direkter Vergleich der Ergebnisse zwischen Citrat und Citrat + Pentadekan wiederum kaum möglich ist.

Bei 2% Citrat und 1% Pentadekan ist immer noch eine deutliche Hemmung des Wachstums zu beobachten, so daß eine vollständige Erholung des Wachstums durch den Citratzusatz nicht eingetreten war. Interessant war die Bestätigung der Ergebnisse in bezug auf die Fettsäuresynthese: Je höher die Ceruleninkonzentration war, desto mehr

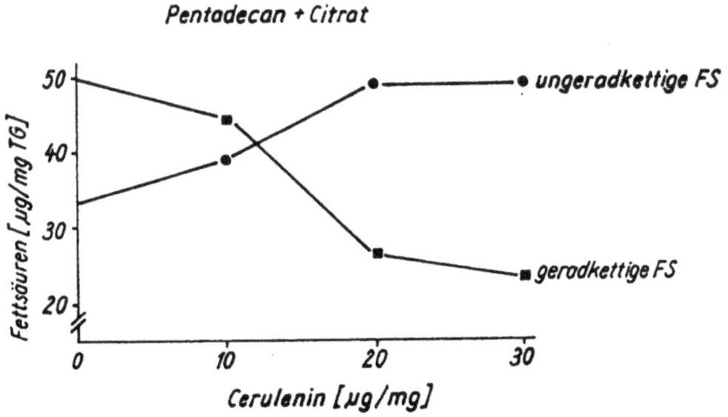

Abb. 4. Bildung von geradkettigen und ungeradkettigen Fettsäuren bei *Mortierella isabellina* in Abhängigkeit von der Cerulenin-Konzentration. Substrat Pentadekan + Citrat

wurde die de novo-Synthese gehemmt. Es traten z. B. immer weniger C_{18}-Säuren auf, und der direkte Einbau von Pentadekansäuren in die Lipide fand vermehrt statt. Die Abb. 4 zeigt die gegenläufige Entwicklung der Bildung von geradkettigen Fettsäuren zu ungeradkettigen Fettsäuren in Abhängigkeit der Ceruleninkonzentration besonders deutlich. Es wurde das gleiche Ergebnis wie bei Verwendung der Alkane als alleinige C-Quelle deutlich. Je stärker die de novo-Fettsäuresynthese gehemmt ist, desto stärker wurde der direkte Einbau der Alkansäuren in die Lipide stimuliert.

IV. Wirkung von Undekansäure auf die Fettsäurebildung

Wir haben in einem breiten „Screening“ die Wirkung verschiedener Fettsäuren, Alkane und Abbauprodukte von Alkanen auf die Regulation der Fettsäurebildung untersucht. Dabei haben sich verschiedene Wirkungen gezeigt, die sich im Fettsäuremuster erkennen lassen.

Es soll im folgenden nur von einem sehr spezifischen Effekt berichtet werden, der von der Undekansäure verursacht wird. Eine ausführliche Beschreibung dieser Wirkung ist an anderer Stelle erfolgt (Hortmann und Rehm [3]).

Im allgemeinen wurde mit 1%iger Undekansäure gearbeitet. Undekansäure ist fest, daher schwimmt sie in glukosehaltigen Lösungen auf der Oberfläche oder ist weitgehend ungelöst im Substrat verteilt. Bei Zusatz von Alkanen wird sie in den Alkanen gelöst, so daß eine homogene lipophile Phase entsteht. Trotz gleicher Einwaagen sind also die effektiven Konzentrationen von Undekansäure je nach Substrat-C-Quelle unterschiedlich.

1. Glukose und Undekansäure

Bei Glukose und Undekansäure tritt zwar eine allgemeine Verminderung des Wachstums im Vergleich zum Wachstum auf Glukose auf, jedoch zeigen sich fast keine Unterschiede im Fettsäuremuster (Tab. 3): Die Fettsäure-de novo-Biosynthese läuft also stark ge-

Tabelle 3. % der Fettsäuren von *Mortierella isabellina* nach Wachstum auf Glukose und n-Alkanen mit und ohne Zusatz von Undekansäure

Fettsäuren	Substrate									
	Glukose	Glukose/ C_{11}-Säure	C_{13}	C_{13}/ C_{11}-Säure	C_{14}	C_{14}/ C_{11}-Säure	C_{15}	C_{15}/ C_{11}-Säure	C_{16}	C_{16}/ C_{11}-Säure
13:0	—	—	6,2	—	—	—	Sp	—	—	—
14:0	1,4	2,2	0,5	0,3	20,6	0,3	Sp	0,2	1,6	0,1
15:0	—	—	10,6	—	—	—	29,3	—	—	—
15:1	—	—	Sp	—	—	—	—	—	—	—
16:0	28,5	31,0	6,8	18,7	13,0	21,7	3,3	20,8	30,4	21,8
16:1	1,9	3,3	0,4	0,4	1,8	0,4	Sp	0,3	2,6	0,6
17:0	—	—	5,1	—	—	—	5,9	—	—	—
17:1	—	—	22,4	—	—	—	28,9	—	—	—
17:2	—	—	3,2	—	—	—	8,1	—	—	—
17:3	—	—	0,4	—	—	—	0,8	—	—	—
18:0	0,2	0,5	Sp	0,3	0,3	0,2	Sp	0,2	0,5	0,2
18:1	49,1	48,9	19,0	51,8	31,8	49,4	10,5	51,4	33,8	47,9
18:2	13,7	11,9	10,6	21,1	16,3	21,4	5,8	20,8	15,8	22,1
18:3	5,3	2,2	13,4	7,4	16,2	7,0	6,0	6,3	15,3	7,3
19:0	—	—	0,1	—	—	—	0,1	—	—	—
19:1	—	—	1,0	—	—	—	1,3	—	—	—
19:2	—	—	—	—	—	—	Sp	—	—	—

hemmt, aber in ihrem Muster ohne wesentlichen Einfluß von Undekansäure ab. Diese wird selbst nicht in die Lipide eingebaut.
Pyruvat- und Acetataufnahme werden durch Undekansäure vollständig gehemmt.

2. Alkane und Undekansäure

Die Tab. 3 zeigt die unterschiedlichen Fettsäuremuster. Bei Anwesenheit von Undekan werden aus Tridekan und auch aus Pentadekan keine ungeradkettigen Fettsäuren gebildet: Es entstehen keine ungeradkettigen Fettsäuren, die dem als Substrat vorgegebenen Alkan entsprechen. Es entstehen auch keine ungeradkettigen Säuren, die aus einer Elongation entstanden sind. Das Fettsäuremuster entspricht bei Anwesenheit von Undekansäure und Alkanen also dem der Glukose, aus der durch die de novo-Synthese nur geradkettige Fettsäuren entstanden sind.

3. Monoterminale Produkte und Undekansäure

Bei Verwendung von Alkoholen der monoterminalen Alkanoxidation als Substrate und Zusatz von Undekansäure liegen im Prinzip die gleichen Ergebnisse wie bei Alkanen als Substrat vor. Kein wesentlicher Einbau der Fettsäure hat stattgefunden, ebenso keine Elongation. Ebenso ist auch der Einfluß der Undekansäure auf Alkansäuren als Substrate.

4. Produkte der subterminalen Alkanoxidation und Undekansäure

Durch die Produkte der subterminalen Alkanoxidation in Gegenwart von Undekansäure wird ein Fettsäuremuster ähnlich dem aus Glukose gebildet. Keine Bildung von ungeradkettigen Fettsäuren ist zu beobachten.

Diskussion

Die Ergebnisse lassen sich folgendermaßen zusammenfassen:

1. Die Fettsäurebildung ist bei den mikrobiellen monoterminalen Oxidationsmechanismen abhängig von der Kettenlänge der Alkane im Substrat.
2. Diese Fettsäuren werden zu einem großen Prozentsatz direkt in die Zellipide eingebaut, zu einem weiteren Teil erfolgt besonders eine β-Oxidation. Dies bedeutet, daß die freien langkettigen Alkansäuren *sowohl* direkt mit dem System II, der langkettigen Acyl-CoA-Synthetase zu Fettsäure-CoA gebunden werden und in den Mitochondrien oder Peroxysomen β-oxidiert werden, *als auch* mit dem System I, der langkettigen Acyl-CoA-Synthetase zur Fettsäure-CoA gebunden werden, um dann evtl. elongiert und mit Glycerin verestert zu werden.
3. Cerulenin hemmt auch beim mycelbildenden Pilz *Mortierella isabellina* die de novo-Fettsäurebiosynthese. Je höher die Ceruleninkonzentrationen sind, um so mehr wird der direkte Einbau von Alkanfettsäuren begünstigt. Durch Cerulenin läßt sich also z. B. eine wesentlich verstärkte Lipidfraktion mit ungeraden Fettsäuren induzieren, wenn man ungeradkettige Alkane vorlegt.
Etwa 15% der de novo-Biosynthese laufen jedoch auch bei Anwesenheit von Cerulenin ab.
4. Cerulenin wirkt nicht auf das alkanoxidierende System, das elongierende System, das desaturierende System (Dies steht im Gegensatz zu Ergebnissen von Buttke und Ingram [1] an *Escherichia coli* K 12), das glycerinesterbildende System.
5. Undekansäure hat eine spezifische Wirkung auf die Elongation von Fettsäuren.

In verschiedenen Versuchen waren die langkettigen Alkane oder ihre primären Oxidationsprodukte die einzige C-Quelle für die Mikroorganismen. Die Ergebnisse zeigten eine funktionierende β-Oxidation auch bei Anwesenheit von Undekansäure. Für die Veresterung der freien Fettsäuren zu ihren CoA-Derivaten sind bei *Candida lipolytica* zwei verschiedene Acyl-CoA-Synthetasesysteme mit unterschiedlichen Funktionen beschrieben worden (KAMIRYO et al. [4]; MISHINA et al. [7]). Das System II befindet sich in den Peroxysomen (MISHINA et al. [8]), wo bei diesem Mikroorganismus auch die β-Oxidation vor sich geht (KAWAMOTO et al. [5]).
TANAKA et al. [13] bestätigten an Mutanten von *C. lipolytica* die beiden Systeme und die Funktion des Systems II bei der β-Oxidation. Das System I ist bei *C. lipolytica* indu-

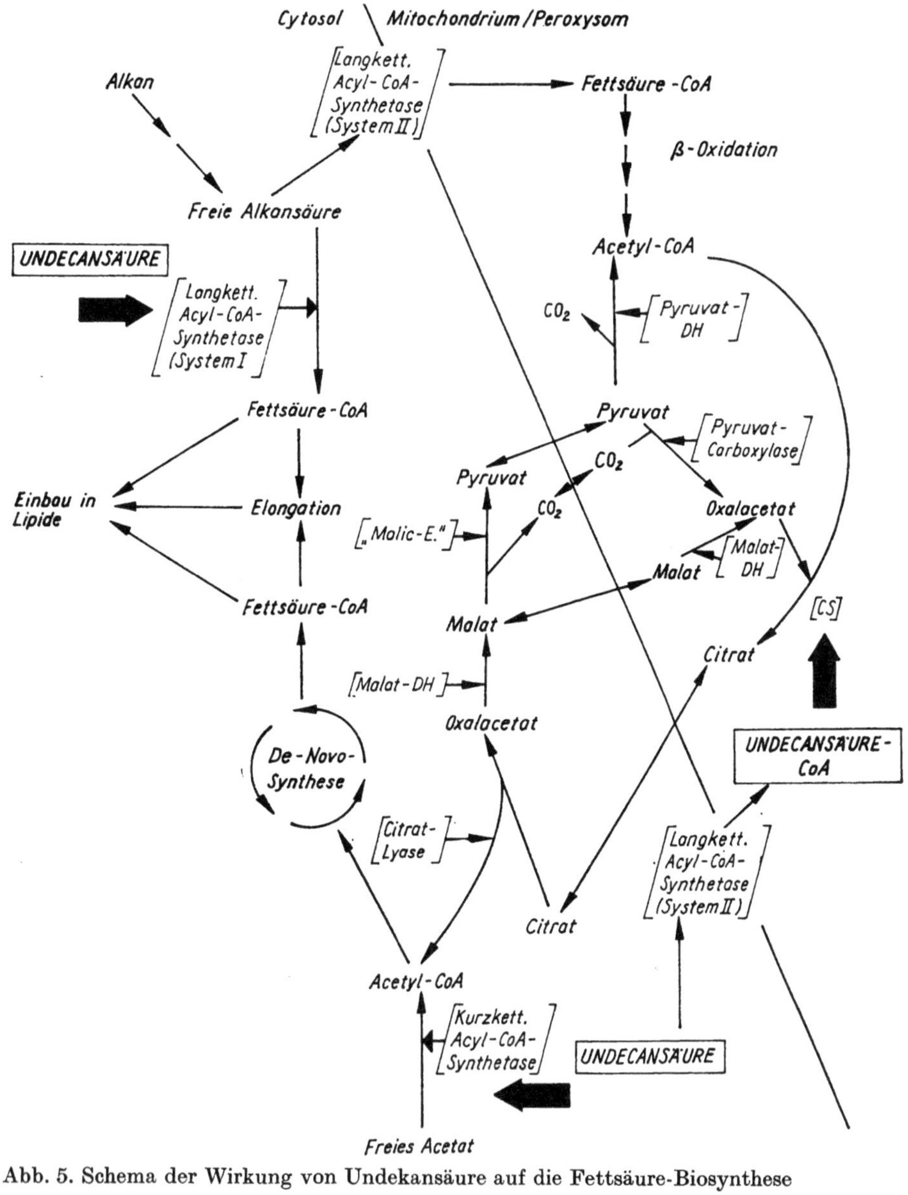

Abb. 5. Schema der Wirkung von Undekansäure auf die Fettsäure-Biosynthese

zierbar und für die Aktivierung der Fettsäuren vor dem Einbau in die zellulären Lipide verantwortlich.

Es ist anzunehmen, daß die Undekansäure bei *Mortierella isabellina* nur die membrangebundene langkettige Acyl-CoA-Synthetase des endoplasmatischen Reticulums (System I) vergiftet. Diese liegt außerhalb der für die β-Oxidation infrage kommenden Zellorganellen. Die freie Undekansäure lag etwa in Konzentrationen von 5 μg/mg TG im Rohextrakt vor. Den möglichen Mechanismus zeigt die Abb. 5.

Auf andere Wirkungen, die in der Abb. 5 angegeben sind, kann hier nicht eingegangen werden: Citrat induziert z. B. bei Alkanansätzen in Anwesenheit von Undekansäure ein sofortiges Wachstum. Acetat wird nicht für die de novo-Biosynthese verwendet, wohl aber Acetyl-CoA.

Inwieweit hier Acetyl-ACP_x bzw. ein Carnitin-shuttle wirksam ist und evtl. durch Undekansäure gehemmt wird, inwieweit evtl. auch ein Acyl-Carrier-Protein (ACP_x) bei der Glycerinveresterung bei *M. isabellina* vorhanden ist und evtl. auch durch Undekansäure gehemmt wird, soll in diesem Beitrag nicht diskutiert werden.

Literatur

[1] BUTTKE, T. M., INGRAM, L. O.: Inhibition of unsaturated fatty acid synthesis in *Escherichia coli* by the antibiotic Cerulenin. Biochemistry **17** (1978) 5282.

[2] HORTMANN, L., REHM, H. J.: Influence of Cerulenin on the fatty acid synthesis of *Mortierella isabellina* grown on long-chain alkanes. Eur. J. Appl. Microbiol. Biotechnol. 1983 (zum Druck gegeben).

[3] HORTMANN, L., REHM, H. J.: Inhibition of fatty acid synthesis in *Mortierella isabellina* by undecane. Eur. J. Appl. Microbiol. Biotechnol. 1983a (zum Druck gegeben).

[4] KAMIRYO, T., MISHINA, M., TASHIRO, S., NUMA, S.: *Candida lipolytica* mutants defective in an acyl-coenzyme A synthetase: isolation and fatty acid metabolism. Proc. Natl. Acad. Sci. USA **74** (1977) 4947.

[5] KAWAMOTO, S., NOZAKI, C., TANAKA, A., FUKUI, S.: Fatty acid β-oxidation system in microbodies of n-alkane-grown *Candida tropicalis*. Eur. J. Biochem. **83** (1978) 609.

[6] KLEBER, H. P., SATTLER, K.: Mikrobiologie und Biochemie der Kohlenwasserstoffwandlung. Leipziger Biotechnologiesymposium 1982, Abstr. der Vorträge und Poster.

[7] MISHINA, M., KAMIRYO, T., TASHIRO, S., HAGIHARA, T., TANAKA, A., FUKUI, S., OSUMI, M., NUMA, S.: Subcellular localization of two long-chain acyl-coenzyme A synthetases in *Candida lipolytica*. Eur. J. Biochem. **82** (1978) 321.

[8] MISHINA, M., KAMIRYO, T., TASHIRO, S., NUMA, S.: Separation and characterization of two long-chain acyl-CoA synthetases from *Candida lipolytica*. Eur. J. Biochem. **82** (1978a) 347.

[9] NOMURA, S. T., HORIUCHI, S., OMURA, S., HATA, T.: The action mechanism of Cerulenin. J. Biochem. (Tokyo) **71** (1972) 783.

[10] REHM, H. J., REIFF, I.: Mechanisms and occurrence of microbial oxidation of long-chain alkanes. Adv. Biochem. Eng. **19** (1981) 175.

[11] REHM, H. J., REIFF, I.: Regulation der mikrobiellen Alkanoxidation mit Hinblick auf die Produktbildung. Acta Biotechnologica **2** (1982) 127.

[12] TANAKA, A., HAGIHARA, T., NISHIKAWA, Y., MISHINA, M., FUKUI, S.: Effects of the antilipogenic antibiotic Cerulenin on growth and fatty acid composition of n-alkane-utilizing *Candida lipolytica*. Eur. J. Appl. Microbiol. **3** (1976) 115.

[13] TANAKA, A., HAGIHARA, T., KAMIRYO, T., MISHINA, M., TASHIRO, S., NUMA, S., FUKUI, S.: *Candida lipolytica* mutants defective in an acyl-CoA synthetase. Fatty acid composition of cells grown on odd-chain n-alkanes. Eur. J. Appl. Microbiol. Biotechnol. **5** (1978) 79.

[14] VANCE, D., GOLDBERG, J., MITSUHASHI, O., BLOCK, K., OMURA, S., NOMURA, S.: Inhibition of fatty acid synthetase by the antibiotic Cerulenin. Biochem. Biophys. Res. Commun. **48** (1972) 649.

[15] YI, Z. H., REHM, H. J.: Metabolic formation of dodecanedioic acid from n-dodecane by a mutant of *Candida tropicalis*. Eur. J. Appl. Microbiol. Biotechnol. **14** (1982) 254.

Acta Biotechnol. 3 (1983) 3, 289–292

Short Communications

Ein Laborkleinfermentorsystem für die biotechnologische Grundlagenforschung — MTA-Ausrüstung

U. Iske[1], H.-J. Rost[1] und S. Portius[2]

[1] Akademie der Wissenschaften der DDR
Institut für technische Chemie Leipzig
7050 Leipzig, Permoserstr. 15

[2] Akademie der Wissenschaften der DDR
Verwaltungs- und Dienstleistungseinrichtungen
7050 Leipzig, Permoserstr. 15

Laborfermentoren stellen ein grundlegendes Arbeitsmittel in der biotechnologischen Forschung dar, dessen Entwicklungsstand in entscheidendem Maße Leistungsfähigkeit und Effektivität der Forschungsprozesse bestimmt. Deshalb ist eng verbunden mit den ständig steigenden Anforderungen an Quantität und Qualität der Informationen aus den verschiedenen hierarchischen Niveaus biotechnologischer Prozesse die Entwicklung hoch instrumentierter leistungsfähiger Laborfermentorsysteme.
Der Trend, biotechnologische Prozesse in ihrer Raum-Zeit-Ausbeute traditionellen chemischen Prozessen anzugleichen, führte zur Entwicklung von Fermentorsystemen mit neuen Wirkprinzipien zur Gewährleistung hoher Stoffübertragungsleistungen. Mit der Realisierung derartiger Hochleistungsfermentoren in großtechnischen Dimensionen ist verbunden die Forderung nach leistungsfähigen Laborfermentoren für die Erarbeitung wissenschaftlicher Grundlagen als Basis für eine gesicherte Maßstabsübertragung.
Die Entwicklung eines 2,5 l Laborsterilfermentors mit hohen Stoffübertragungsleistungen stellt einen Beitrag zur Vervollständigung der gerätetechnischen Basis für die biotechnologische Grundlagenforschung dar.

Technische Beschreibung

Als Reaktorgefäß wurde eine Glas-Stahl-Kombination mit einem Gesamtvolumen von 2,5 l entsprechend einem Arbeitsvolumen zwischen 600 und 1500 ml gewählt. Diese Kombination von Glaszylinder und Stahlring gestattet neben der ungestörten optischen Kontrolle des Fermentationsprozesses den problemlosen Einbau einer größeren Anzahl von Meßsonden ohne signifikante Beeinflussung der hydrodynamischen Verhältnisse im Reaktor. Durch spezielle Elektrodenaufnehmer wird der wahlweise Einsatz unterschiedlicher Elektrodensysteme ermöglicht, z. B.

pH-Elektrode	Typ EGC 250.31, Typ EGA 80 Forschungsinstitut Meinsberg/Sa.
pO_2-Elektrode	Typ SMZ 300 VEB Metra, Radebeul Typ ME 22/23 Forschungsinstitut Meinsberg/Sa.

	Galvanische Meßsonde
	Institut für technische Chemie, AdW der DDR
pCO_2-Elektrode	Forschungsinstitut Meinsberg/Sa.

Das Reaktorgefäß einschließlich seines peripheren Zubehörs wie Filter und Dosierleitungen ist in einem ortsveränderlichen standsicheren Gestell untergebracht und kann durch mechanische Höhenverstellung in eine optimale Arbeitsposition gebracht werden.
In Abb. 1 ist der Aufbau des Fermentorsystems dargestellt. Als Rührsystem wird ein 5-Blattscheibenrührer verwendet, dessen Antrieb durch einen 50 W Wechselstrommotor maximaler Drehzahl 3000 U/min (begrenzt auf 1500 U/min) über eine Magnetkupplung (Ringmagnete, Werkstoff Manipern 860, 6fach polarisiert) vom Reaktordeckel aus erfolgt. Neben dem Blattrührer trägt die Rührerwelle in ihrem oberen Drittel einen mechanischen Schaumzerstörer in Form eines achtflügeligen Propellerrades. Beide Funktionselemente sind in ihrer Höhe verstellbar angeordnet.

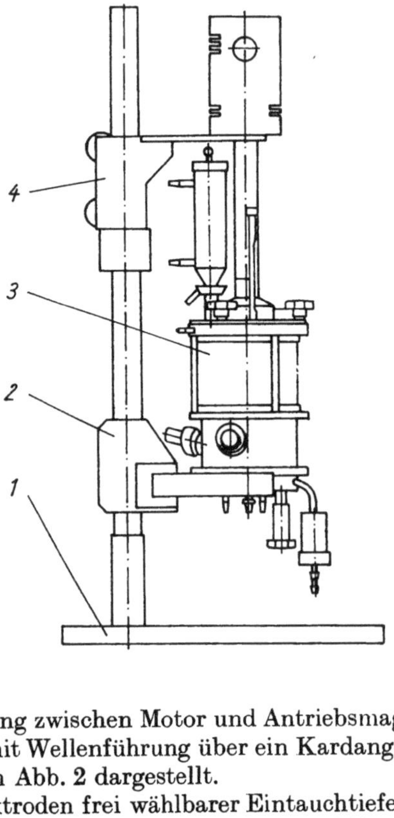

Abb. 1. Gesamtansicht Fermentorsystem

1 Grundplatte mit Säule, 2 Fermentorhalterung (verstellbar), 3 Reaktorgefäß mit Funktionselementen, 4 Motorhalterung (verstellbar)

Die Kopplung zwischen Motor und Antriebsmagnet erfolgt über eine elastische Klauenkupplung mit Wellenführung über ein Kardangelenk. Das Antriebssystem mit Reaktordeckel ist in Abb. 2 dargestellt.
Schaumelektroden frei wählbarer Eintauchtiefen am Reaktordeckel können als Signalgeber für eine Schaumregulierung mit Hilfe chemischer Agentien eingesetzt werden. Zur Dosierung flüssiger Substrate sind 4 Dosierleitungen im oberen Stahlring vorgesehen, die bei Nichtbenutzung von innen mit speziellen Gummikappen verschlossen werden können.
Die Belüftung des Fermentationsmediums erfolgt über austauschbare Begasungselemente (Düse, Lochplatte). Zur Erzeugung hoher Turbulenzen als Voraussetzung für einen effektiven Stoffübergang dient ein Bufflesystem, das gleichzeitig die Funktion des Kühlsystems übernimmt.

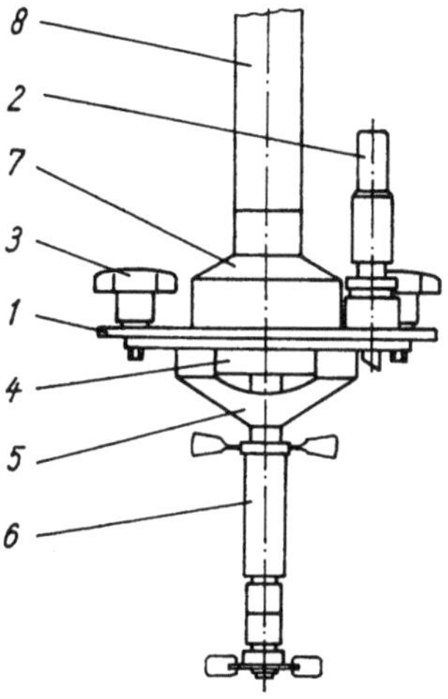

Abb. 2. Reaktordeckel mit Antrieb

1 Deckelplatte, 2 Einfüllstutzen, NS 12/21, 3 Deckelverschraubung, 4 Innenmagnet (gekapselt), 5 Innenlagerung, 6 Rührerwelle mit Schaumbrecher und Blattrührer, 7 Antriebsmagnet mit Außenlagerung (abnehmbar), 8 Antriebswelle mit Klauenkupplung verkleidet

Die Kopplung aller Ver- bzw. Entsorgungsleitungen erfolgt am Reaktor über Normschliffverbindungen.
Die Massen- bzw. Volumenkonstanz bei kontinuierlicher Betriebsweise ist sowohl mit Hilfe austauschbarer Überlaufrohre verschiedener Abmessungen als auch durch Ansteuerung eines sterilisierbaren Membranventils in einer gesonderten Ernteleitung realisierbar. Ein dampfsterilisierbares Ventil im Reaktorboden gestattet eine kontaminationsfreie Probenahme während des Prozeßverlaufes. Abb. 3 zeigt den Fermentorboden einschließlich seiner funktionellen Baugruppen.

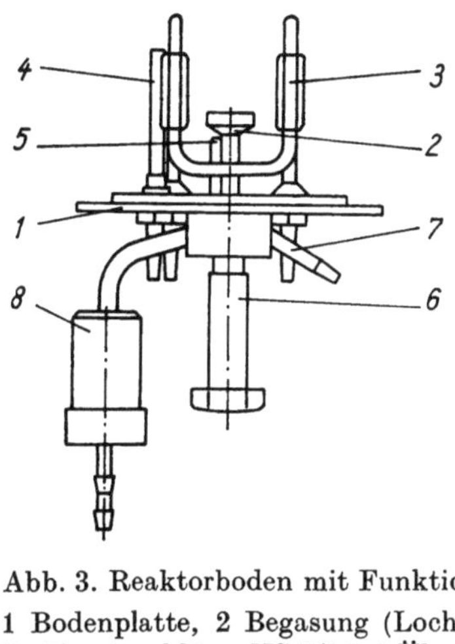

Abb. 3. Reaktorboden mit Funktionselementen

1 Bodenplatte, 2 Begasung (Lochplatte), Anschluß NS 7/16, 3 Kühlschlange mit Buffle, Anschlüsse NS 7/16, 4 Überlaufrohr, Anschluß NS 10/16, 5 Hülse mit Widerstandsthermometer, 6 Probenahmeventil, 7 Dampfanschluß NS 7/16, 8 Probenehmer

Zur Kühlung des Abgasstromes wird ein Rohrbündelkühler mit integriertem Sterilfilter (Spezialfilter NKF 500, VEB Feinpapierfabriken Neu Kaliß) eingesetzt, Abb. 4.
Die Sterilisation des Reaktorgefäßes einschließlich seines peripheren Zubehörs erfolgt nach Entfernung des Magnetantriebes am Reaktordeckel im Autoklaven.
Nach Lösen von 2 Arretierungsschrauben kann das Reaktorgefäß problemlos aus dem Gestell entnommen werden.

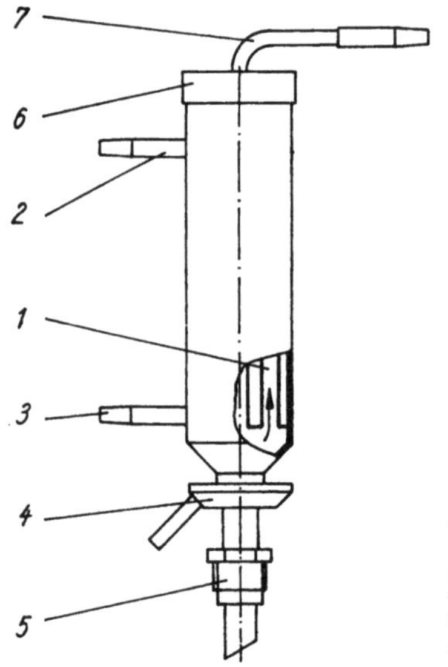

Abb. 4. Abgaskühler
1 Rohrbündelkühler, 2 Kühlwassereingang NS 7/16, 3 Kühlwasserausgang NS 7/16, 4 Kondensatrinne mit Ablauf, 5 Deckelanschluß, 6 Überwurf mit Filtereinlage, 7 Abgasleitung

Bestimmung der Sauerstoffeintragsleistung

Die Ermittlung der Sauerstoffeintragsleistung erfolgte nach der Sulfitmethode als Funktion von Rührerdrehzahl, Begasungsintensität und Begasungssystem.
Der maximal erreichte Sauerstoffeintrag von 6,6 g/l · h bei 1500 U/min und einer Begasung von 120 l/h kann für Fermentoren dieser Baugröße als sehr gut eingeschätzt werden. Interessanterweise ermöglicht die Düsenbegasung bei hohen Drehzahlen einen höheren Sauerstoffübergang als die Lochplatte, die ihrerseits wiederum effektiver bei niedrigeren Drehzahlen arbeitet.

Eingegangen: 28. 1. 1982

Acta Biotechnol. 3 (1983) 3, 293–296

Morphometrische Verfolgung der enzymatischen Hydrolyse von Cellulose

D. C. Dan[1], K.-J. Linow[1], B. Philipp[1], G. Schulz[2], E. W. Unger[3]

[1] Institut für Polymerenchemie der Akademie der Wissenschaften der DDR, 153 Teltow-Seehof, Kantstr. 55
[2] Humboldt-Universität Berlin, Sektion Nahrungsgüterwirtschaft
[3] TU Dresden, Sektion Verarbeitungs- und Verfahrenstechnik

In früheren Publikationen [1, 2] hatten wir die morphologischen Veränderungen bei der enzymatischen Hydrolyse unterschiedlich vorbehandelter Linterscellulose beschrieben, wobei die Oberflächenstruktur und die Partikellänge bzw. -längenverteilung vor und nach dem enzymatischen Abbau betrachtet und auf einen ausgeprägten Querzerfall der Fasern hingewiesen wurde. Nachdem heute mit der von Unger [3] entwickelten Anordnung zur streulichtfotometrischen Faserlängenmessung in einer strömenden Stoffsuspension ein Meßsystem zur Verfügung steht, das eine objektive, vollautomatisierte, rechnergestützte Längenklassierung von mehreren 1000 Partikeln pro Minute gestattet, erschien es aussichtsreich, mit dieser Apparatur die Möglichkeit einer unmittelbaren morphometrischen Verfolgung des zeitlichen Ablaufs heterogener Hydrolyseprozesse der Cellulose zu prüfen.

Als Modellsubstrat verwendeten wir einen Buchensulfit-Textilzellstoff im initialfeuchten und im bis Raumtemperatur an der Luft getrockneten Zustande. Als Hydrolysemedien dienten auf pH = 5 gepufferte Kulturfiltrate von *Gliocladium spec.* HUG k 2642 [4] und 1 N HCl, wobei in allen Fällen bei 40 °C ± 1 °C gearbeitet wurde. Das Substrat wurde als klumpenfreie wäßrige Stoffsuspension dem umlaufenden Hydrolysemedium zugesetzt, wobei die Konzentration 10 mg Trockensubstanz in 500 ml Medium betrug. Nach wenigen Minuten lag eine gleichmäßige Verteilung der Fasern in dem in einem geschlossenen Kreislauf geförderten Medium vor, so daß nach spätestens 5 Minuten mit der Verfolgung des Hydrolyseverlaufes begonnen werden konnte. In Abständen von 5 bzw. 10 min wurde jeweils während 25,6 s eine Zählung und Klassierung der Partikel in einem Durchflußvolumen von 1,6 ml vorgenommen. Blindwerte für das substratfreie Medium wurden separat ermittelt.

Abb. 1 veranschaulicht die Änderung der auf einen Anfangswert bezogenen Gesamtpartikelzahl während jeweils einer Zählperiode einerseits beim enzymatischen, andererseits beim säurehydrolytischen Abbau des initialfeuchten Buchenzellstoffs. Abb. 2 gibt einen Vergleich von initialfeuchtem und lufttrockenem Stoff. Zur Linearisierung und quantitativen Auswertung der Kurvenverläufe eignet sich nach unseren Erfahrungen ein zweiparametriges Zeitgesetz erster Ordnung

$$-\frac{dZ}{dt} = k_1(Z - Z_\infty)$$

wobei Z die jeweilige Gesamtpartikelzahl, k_1 eine Geschwindigkeitskonstante 1. Ordnung und Z_∞ die über ein „Anpassungsprogramm“ [5] rechnerisch ermittelte Anzahl „unlöslicher“ Partikel darstellt, die annähernd der nach langer Reaktionszeit gefun-

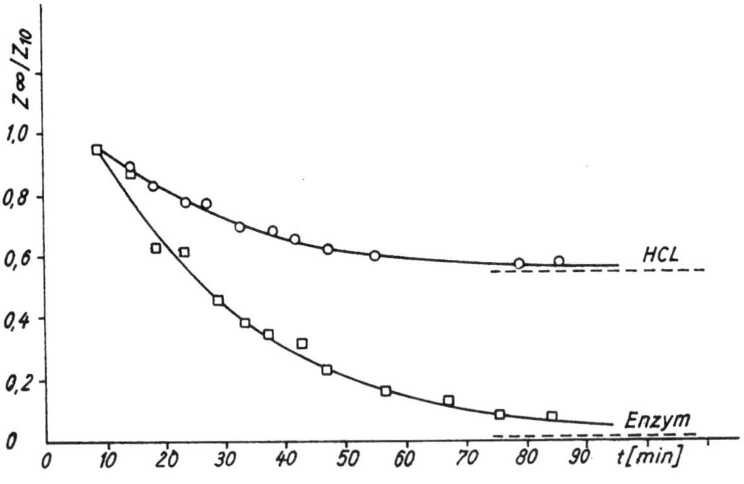

Abb. 1. Veränderung der relativen Gesamtpartikelzahl im Verlaufe des enzymatischen und säurehydrolytischen Abbaues von initialfeuchtem Buchensulfit-Textilzellstoff

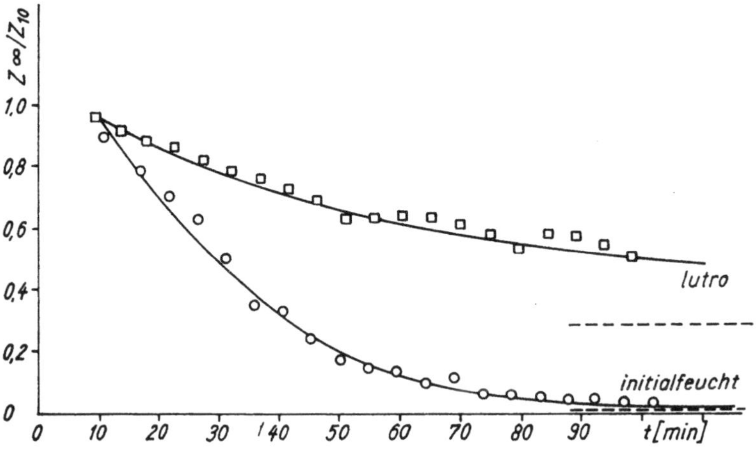

Abb. 2. Veränderung der relativen Gesamtpartikelzahl im Verlaufe des hydrolytischen Abbaues von initialfeuchtem und lufttrockenem Buchensulfit-Textilzellstoff

denen Partikelzahl entspricht. Tabelle 1 gibt die entsprechenden Zahlenwerte für einige Versuchsreihen wieder, wobei die Unterschiede zwischen enzymatischem und säurehydrolytischem Abbau einerseits, initialfeuchtem und lufttrockenem Stoff andererseits deutlich in Erscheinung treten. Für enzymatische Hydrolyse und Säurehydrolyse werden unter den hier gewählten Bedingungen annähernd gleiche k_1-Werte erhalten; die Unterschiede im Kurvenverlauf spiegeln sich jedoch in den sehr unterschiedlichen „Restwerten“ Z_∞/Z_{10} deutlich wider.

Die zeitliche Änderung der Besetzung der einzelnen Partikellängenklassen beim enzymatischen und beim säurehydrolytischen Abbau des initialfeuchten Buchenzellstoffs geht aus den Tabellen 2 und 3 hervor. In beiden Fällen spiegelt sich ein rascher

Tabelle 1. Kinetische Kenngrößen für die zeitliche Abnahme der Gesamtpartikelzahl bei der heterogenen Hydrolyse von Buchensulfitzellstoff

Medium	Substrat	$10^2\,k_1$ [min^{-1}]	Z_∞/Z_0 [%]*
1 n HCl/40 °C	initialfeucht	4,0	58
Enzymlösung	initialfeucht	3,6	< 1
Enzymlösung	lufttrocken	1,1	31
Enzymlösung	initialfeucht	3,8	< 1

* Zur besseren Vergleichbarkeit wurden die Angaben für Z_∞ auf einen „Anfangswert" Z_{10} nach 10 Minuten Reaktionszeit bezogen.

Tabelle 2. Änderung der Partikellängenverteilung beim säurehydrolytischen Abbau eines initialfeuchten Buchenzellstoffes

t [min]	$\sum Zi$	Partikelzahl Zi in Größenklassen 3	4	5	6	7	8	9
vor der Hydrolyse	3960	0	28	424	1922	1315	164	104
10	2372	0	16	264	1194	868	28	0
15	2239	0	16	261	1182	779	0	0
20	2096	0	22	230	1152	668	29	0
25	1948	0	12	211	1028	678	19	0
30	1938	0	12	200	974	704	46	0
35	1747	0	11	201	961	573	0	0
40	1725	0	9	198	938	579	0	0
45	1653	0	11	190	920	530	0	0
50	1570	0	11	163	910	485	0	0
60	1517	0	8	188	794	506	19	0
70	—	—	—	—	—	—	—	—
80	1432	0	6	132	740	525	26	0
90	1452	0	6	128	765	548	0	0

„Querzerfall" der Fasern im völligen Verschwinden der „langen" Partikel (Klasse 9) binnen weniger Minuten wider. Beim enzymatischen Abbau setzt sich dieser Prozeß insofern fort, als nach längerer Reaktionsdauer die 4 oberen Längenklassen „leer" bleiben, während in der untersten besetzten Klasse (Klasse 4) die Partikelzahl über einen längeren Zeitraum annähernd konstant bleibt. Während im Falle des enzymatischen Abbaues nach langer Reaktionszeit nur noch 2 der zur Klassierung verfügbaren 7 Klassen besetzt sind, ist im Falle der Säurehydrolyse unter den hier angewandten Bedingungen am Bu—Si-Zellstoff eine wesentlich breitere Verteilung der Partikellängen (4 bzw. 5 besetzte Klassen) festzustellen.

Voraussetzungen für die erfolgreiche Anwendung der hier beschriebenen Arbeitsweise sind einmal die getrennte Erfassung und ggf. Berücksichtigung der im Abbaumedium vorhandenen Inhomogenitäten (Partikel und Luftbläschen) auch hinsichtlich ihrer zeitlichen Anzahl- und Größenänderung, sowie das Vorliegen einer von Agglomeraten („Knoten") freien Substratsuspension. Weiterhin ist zu berücksichtigen, daß die untere Grenze der Erfaßbarkeit für die derzeitige Meßanordnung bei einer Partikellänge von etwa 5 µm liegt und somit über das Vorliegen noch kleinerer Faserfragmente

keine Aussagen getroffen werden können. Bei Erfüllung der oben genannten Prämissen erscheint die hier vorgeschlagene Arbeitsweise geeignet, die zeitliche Änderung von Partikelzahl, -größe und -größenverteilung vor allem im Anfangsstadium der heterogenen Hydrolyse von Cellulosefasersubstraten im Echtzeitbetrieb zu erfassen und eine quantitative kinetische Beschreibung der auftretenden morphologischen Veränderungen zu ermöglichen.

Tabelle 3. Änderung der Partikellängenverteilung beim enzymatischen Abbau eines initialfeuchten Buchenzellstoffes

t [min]	ΣZi	Partikelzahl Zi in Größenklassen						
		3	4	5	6	7	8	9
vor der Hydrolyse z. Vergl.	3960	0	28	424	1922	1315	164	104
10	1760	0	33	312	1054	322	0	0
15	1637	0	26	298	967	343	0	0
20	1167	0	28	324	817	0	0	0
25	1149	0	25	313	809	0	0	0
30	852	0	20	229	596	0	0	0
35	703	0	26	256	486	0	0	0
40	637	0	28	294	315	0	0	0
45	585	0	22	229	331	0	0	0
50	427	0	27	178	221	0	0	0
60	280	0	24	165	90	0	0	0
70	225	0	22	203	0	0	0	0
80	140	0	23	117	0	0	0	0
90	129	0	14	115	0	0	0	0

Literatur

[1] Philipp, B., Dan, D. C., Schaper, A., Unger, E. W.: Cell. Chem. Technol. (im Druck).
[2] Philipp, B., Dan, D. C., Fink, H.-P.: Ekman-Symposion Stockholm 1981, Tagungsband IV, 79.
[3] Philipp, B., Schlacken, H., Stiller, H. J., Unger, E. W.: Paperi ja Puu **60** (1978) 681.
[4] Hirte, W. F., Glathe, I., Gradel, A., Schade, W., Huber, T., DD-WP 149914.
[5] Philipp, B., Dan, D. C., Linow, K. J., Polter, E., Schulz, G., Acta Biotechnol. **2** (1982) 275.

Microbes and Engineering Aspects

Herausgegeben von A. Fiechter

1982. 224 Seiten — 28 Abbildungen — gr. 8° — 92,— M
Bestell-Nr. 763 070 4 (6680)

Das Buch enthält Beiträge zur mikrobiologischen Biomasseproduktion von Methanol und Äthanol. Verfügbare Substrate, Prozeßalternativen und künftige Möglichkeiten werden beschrieben und diskutiert. Außerdem wird ein einfach strukturiertes Modell zur Beschreibung des Primärstoffwechsels für die technische Mikrobiologie vorgestellt. Energetische Fragen der Thermodynamik, der ATP-Bedarf und biotechnische Fragen bei der Bindung von Stickstoff durch Mikroorganismen sind Inhalt eines weiteren Beitrages.

Bestellungen durch eine Buchhandlung erbeten

A K A D E M I E - V E R L A G

DDR - 1086 Berlin, Leipziger Straße 3—4

Acta Biotechnologica

Volume 3 1983 Number 3

Contents

Short Communications

Acta Biotechnologica is indexed in Current Contents/ET & AT;
Chemical Abstracts; Biotechnology Abstracts.

Bei Fragen zur Produktsicherheit wenden Sie sich bitte an:
If you have any questions regarding product safety,
please contact:

Walter de Gruyter GmbH
Genthiner Straße 13
10785 Berlin
productsafety@degruyterbrill.com

AF877051